T0142496

Adaptation, Learning, and Optimization

Volume 20

Series editors

Meng-Hiot Lim, Nanyang Technological University, Singapore
e-mail: emhlim@ntu.edu.sg

Yew Soon Ong, Nanyang Technological University, Singapore
e-mail: asysong@ntu.edu.sg

About this Series

The role of adaptation, learning and optimization are becoming increasingly essential and intertwined. The capability of a system to adapt either through modification of its physiological structure or via some revalidation process of internal mechanisms that directly dictate the response or behavior is crucial in many real world applications. Optimization lies at the heart of most machine learning approaches while learning and optimization are two primary means to effect adaptation in various forms. They usually involve computational processes incorporated within the system that trigger parametric updating and knowledge or model enhancement, giving rise to progressive improvement. This book series serves as a channel to consolidate work related to topics linked to adaptation, learning and optimization in systems and structures. Topics covered under this series include:

- complex adaptive systems including evolutionary computation, memetic computing, swarm intelligence, neural networks, fuzzy systems, tabu search, simulated annealing, etc.
- machine learning, data mining & mathematical programming
- hybridization of techniques that span across artificial intelligence and computational intelligence for synergistic alliance of strategies for problem-solving.
- aspects of adaptation in robotics
- agent-based computing
- autonomic/pervasive computing
- dynamic optimization/learning in noisy and uncertain environment
- systemic alliance of stochastic and conventional search techniques
- all aspects of adaptations in man-machine systems.

This book series bridges the dichotomy of modern and conventional mathematical and heuristic/meta-heuristics approaches to bring about effective adaptation, learning and optimization. It propels the maxim that the old and the new can come together and be combined synergistically to scale new heights in problem-solving. To reach such a level, numerous research issues will emerge and researchers will find the book series a convenient medium to track the progresses made.

More information about this series at http://www.springer.com/series/8335

Slim Bechikh · Rituparna Datta
Abhishek Gupta
Editors

Recent Advances in Evolutionary Multi-objective Optimization

Editors
Slim Bechikh
SOIE lab, Computer Science Department
University of Tunis, ISG-Tunis
Tunis
Tunisia

Abhishek Gupta
School of Computer Engineering
Nanyang Technological University
Singapore
Singapore

Rituparna Datta
Graduate School of Knowledge Service Engineering, Department of Industrial and Systems Engineering
Korean Advanced Institute of Science and Technology
Daejeon
Republic of Korea

and

Department of Mechanical Engineering
Indian Institute of Technology Kanpur
Kanpur, Uttar Pradesh
India

ISSN 1867-4534 ISSN 1867-4542 (electronic)
Adaptation, Learning, and Optimization
ISBN 978-3-319-82709-4 ISBN 978-3-319-42978-6 (eBook)
DOI 10.1007/978-3-319-42978-6

Printed on acid-free paper

This Springer imprint is published by Springer Nature
The registered company is Springer International Publishing AG Switzerland

To my family

—Slim Bechikh

To my spouse Anima

—Rituparna Datta

To my parents

—Abhishek Gupta

Preface

This book surveys the recent advances in the field of evolutionary multi-objective optimization. In fact, most real problems are multi-objective in nature, i.e. they involve multiple conflicting objectives to be minimized or maximized simultaneously in limited resources. The resolution of such type of problems gives rise to a set of non-dominated solutions forming the Pareto front. Evolutionary algorithms have been recognized to be well-suited to solve multi-objective problems, thanks to their ability in providing the decision-maker with a set of trade-off solutions in a single run in addition to their insensitivity to the geometrical features of the objective space. However, real-world applications usually have one or several aspects that need further efforts to be tackled. In this book, we survey recent achievements in handling five aspects. The first aspect is dynamicity where the objective functions and/or the constraints may change over time. In this case, the optimization algorithm should track the Pareto front after the occurrence of any change. The second aspect is the presence of hierarchy between the objectives. This kind of problems is called bi-level where an upper level problem has a lower level one in its constraints. The main difficulty in bi-level programming is that the evaluation of an upper level solution requires finding the optimal lower level one, which is computationally expensive. The third aspect is the objective space high dimensionality. This aspect means solving many-objective problems involving more than three objectives. The main difficulty in dealing with such type of problems is that most solutions become equivalent to each others; therefore making the algorithm behaving like random search. The fourth aspect is the emerging notion of evolutionary multitasking which is inspired by the cognitive ability to multitask. Shown to be a natural extension of population-based search algorithms, multitasking encourages multiple heterogeneous search spaces belonging to distinct tasks to be unified and searched concurrently. The resultant knowledge exchange provides the scope for improved convergence characteristics across multiple tasks at once, thereby facilitating enhanced productivity in decision-making processes. The fifth aspect is the presence of constraints where the evolutionary algorithm should search for solutions in the decision space while respecting a set of predefined constraints so that it outputs a set of feasible non-dominated solutions.

This book provides both methodological treatments and real-world insights gained by experience, all contributed by specialized researchers. As such, it is a comprehensive reference for researchers, practitioners, and advanced-level students interested in both the theory and the practice of using evolutionary algorithms in tackling real-world applications involving multiple objectives. The book provides a comprehensive treatment of the field by offering chapters whose topics are disjoint or having minimal overlaps, each tackling a single multi-objective aspect. Moreover, the last chapter highlights a number of practical applications showing the usability of multi-objective evolutionary algorithms in practice; thereby motivating researchers and engineers to use evolutionary approaches in solving their encountered problems.

Tunis, Tunisia — Slim Bechikh
Daejeon, Republic of Korea — Rituparna Datta
Singapore, Singapore — Abhishek Gupta

Contents

About the Editors

Dr. Slim Bechikh received the B.Sc., the M.Sc., the Ph.D., and the Habilitation degrees in Computer Science with Business from the University of Tunis (ISG), Tunis, Tunisia, in 2006, 2008, 2013, and 2015 respectively. He is Associate Professor with the Computer Science Department of the University of Carthage (FSEG), Nabeul, Tunisia. His main research interests include multi-objective optimization, evolutionary computation, bi-level programming, and their applications. He worked as a researcher for 5 years within the Optimization Strategy and Intelligent Computing lab (SOIE), Tunisia. He is a reviewer for many journals such as IEEE Transactions on Evolutionary Computation, IEEE Transactions on Cybernetics, Soft Computing, etc; and many conferences such as IEEE CEC, ACM GECCO, ACM SAC, etc. More information about his research can be found from his Webpage: https://sites.google.com/site/slimbechikh/. E-mail: slim.bechikh@gmail.com

Dr. Rituparna Datta is a postdoctoral researcher with Graduate School of Knowledge Service Engineering, Department of Industrial Systems Engineering, Korea Advanced Institute of Science and Technology (KAIST), Republic of Korea. Prior to that, he was a postdoctoral fellow in INRIA, France, Project scientist at SMSS lab, IIT Kanpur, India and postdoctoral research fellow at the Robot Intelligence Technology (RIT) Laboratory, Korea Advanced Institute of Science and Technology (KAIST). He earned his Ph.D. in Mechanical Engineering at Indian Institute of Technology (IIT) Kanpur in 2013. His current research work involves investigation of Evolutionary Algorithms-based approaches to constrained optimization, applying multi-objective optimization in engineering design problems, surrogate-assisted optimization, memetic algorithms, derivative-free optimization, and robotics. He is a member of ACM, IEEE, and IEEE Computational Intelligence Society. He has been invited to deliver lectures at several institutes and universities across the globe, including at the Trinity College Dublin (TCD), Delft University of Technology (TUDELFT), University of Western Australia (UWA), University of Minho, Portugal, University of Nova de Lisboa, Portugal, University of Coimbra, Portugal, and IIT Kanpur, India. He is a regular reviewer of IEEE

Transactions on Evolutionary Computation, Journal of Applied Soft Computing, Journal of Engineering Optimization, Journal of The Franklin Institute, and International Journal of Computer Systems in Science and Engineering, and was in the program committee of Genetic and Evolutionary Computation Conference (GECCO 2014), iNaCoMM2013, GECCO 2013, GECCO 2012, GECCO 2011, eighth international conference on Simulated Evolution And Learning (SEAL 2010), International Conference on Molecules to Materials (ICMM-06), and some Indian conferences. He has also chaired a session in ACODS 2014 and UKIERI Workshop on Structural Health Monitoring 2012, GECCO 2011, IICAI 2011, to name a few. E-mail: rdatta@kaist.ac.kr and rdatta@iitk.ac.in

Dr. Abhishek Gupta is a Research Fellow at the Rolls-Royce Corporate Lab at Nanyang Technological University, Singapore. He earned his Ph.D. in Engineering Science from the University of Auckland, New Zealand, in 2014, working on the numerical modelling and optimization of non-isothermal fluid flows in porous media, with application to composites manufacturing processes. He received a bachelors degree from the National Institute of Technology (NIT) Rourkela, India in 2010. His research interests lie in the field of computational science, spanning topics in continuum mechanics as well as computational intelligence. His most recent research activities have primarily been in evolutionary computation, with particular emphasis on multi-objective bi-level programming and multitasking in optimization. He is a member of IEEE Computational Intelligence Society's Pre-College Activities Committee and serves as a reviewer for the IEEE Transactions on Evolutionary Computation. E-mail: abhishekg@ntu.edu.sg

Multi-objective Optimization: Classical and Evolutionary Approaches

Maha Elarbi, Slim Bechikh, Lamjed Ben Said and Rituparna Datta

Abstract Problems involving multiple conflicting objectives arise in most real world optimization problems. Evolutionary Algorithms (EAs) have gained a wide interest and success in solving problems of this nature for two main reasons: (1) EAs allow finding several members of the Pareto optimal set in a single run of the algorithm and (2) EAs are less susceptible to the shape of the Pareto front. Thus, Multi-objective EAs (MOEAs) have often been used to solve Multi-objective Problems (MOPs). This chapter aims to summarize the efforts of various researchers algorithmic processes for MOEAs in an attempt to provide a review of the use and the evolution of the field. Hence, some basic concepts and a summary of the main MOEAs are provided. We also propose a classification of the existing MOEAs in order to encourage researchers to continue shaping the field. Furthermore, we suggest a classification of the most popular performance indicators that have been used to evaluate the performance of MOEAs.

Keywords Multi-objective optimization · Evolutionary algorithms · Test problems · Performance metrics

M. Elarbi (✉) · S. Bechikh · L. Ben Said
SOIE Lab, Computer Science Department, ISG-Tunis, University of Tunis,
Bouchoucha City, 2000 Le Bardo, Tunis, Tunisia
e-mail: arbi.maha@yahoo.com

S. Bechikh
e-mail: slim.bechikh@gmail.com

L. Ben Said
e-mail: lamjed.bensaid@isg.rnu.tn

R. Datta
Graduate School of Knowledge Service Engineering,
Department of Industrial and Systems Engineering,
Korean Advanced Institute of Science and Technology (KAIST), 291 Daehak-ro,
Yuseong-gu, Daejeon 34141, Republic of Korea
e-mail: rdatta@kaist.ac.kr; rdatta@iitk.ac.in

R. Datta
Department of Mechanical Engineering,
Indian Institute of Technology Kanpur, Kanpur, Uttar Pradesh, India

S. Bechikh et al. (eds.), *Recent Advances in Evolutionary Multi-objective Optimization*, Adaptation, Learning, and Optimization 20,
DOI 10.1007/978-3-319-42978-6_1

1 Introduction

Most real world optimization problems involve the optimization of two or more conflicting objectives simultaneously. In order to solve a MOP, there are three goals to pursue: (1) convergence, (2) diversity, and (3) solution distribution uniformity. In fact, the obtained non-dominated solutions should be as close as possible to the Pareto optimal front of the optimization problem. This goal is similar to the demand of convergence to the global optimum in single-objective optimization. Often, there exist an infinite number of Pareto optimal solutions. Naturally, only a finite number of solutions can be generated during an optimization process. Furthermore, the number of generated solutions must be limited otherwise the computational cost would become too large. Nevertheless, the largest possible freedom of choice should be offered to the Decision Maker (DM). Therefore, a well-distributed approximation set is demanded which is a goal that consists itself of two requirements: (1) an extent that is as large as possible and (2) a distribution that is as evenly spaced as possible. Pareto optimal fronts may be disconnected, so in that case an exactly uniform distribution of solutions is not possible. Nevertheless, the non-dominated solutions should cover all regions of the Pareto-optimal front and reproduce the curvature of the underlying Pareto optimal front as correctly as possible. These demands do not have a counterpart in single-objective optimization since in that case only one solution is generated.

A MOP consists in minimizing or maximizing an objective function vector under some constraints. The general form of a MOP is as follows [1]:

$$\begin{cases} Min f(x) = [f_1(x), f_2(x), \ldots, f_M(x)]^T & \\ g_j(x) \geq 0 & j = 1, \ldots, P \\ h_k(x) = 0 & k = 1, \ldots, Q \\ x_i^L \leq x_i \leq x_i^U & i = 1, \ldots, n \end{cases} \quad (1)$$

where M is the number of objective functions, P is the number of inequality constraints, Q is the number of equality constraints, x_i^L and x_i^U correspond respectively to the lower and upper bounds of the variable (This notation is assumed throughout the overall chapter). A solution x_i satisfying the $(P+Q)$ constraints is said feasible and the set of all feasible solutions defines the feasible search space denoted by Ω. In this formulation, we consider a minimization MOP since maximization can be easily turned to minimization based on the duality principle by multiplying each objective function by -1 and transforming constraints based on the duality rules.

The resolution of a MOP yields a set of trade-off solutions, called Pareto optimal solutions or non-dominated solutions, and the image of this set in the objective space is called the Pareto front. Hence, the resolution of a MOP consists in approximating the whole Pareto front. In the following, we give some background definitions related to multi-objective optimization:

Definition 1 (*Pareto optimality*) A solution $x^* \in \Omega$ is Pareto optimal if $\forall x \in \Omega$ and $I = \{1, \ldots, M\}$ either $\forall m \in I$ we have $f_m(x) = f_m(x^*)$ or there is at least one $m \in I$ such that $f_m(x) > f_m(x^*)$. The definition of Pareto optimality states that x^*

is Pareto optimal if no feasible vector x exists which would improve some objectives without causing a simultaneous worsening in at least another one.

Definition 2 (*Pareto dominance*) A solution $u = (u_1, u_2, \ldots, u_n)$ is said to dominate another solution $v = (v_1, v_2, \ldots, v_n)$ (denoted by $f(u) \prec f(v)$) if and only if *f(u)* is partially less than *f(v)*. In other words, $\forall m \in \{1, \ldots, M\}$ we have $f_m(u) \leq f_m(v)$ and $\exists m \in \{1, \ldots, M\}$ where $f_m(u) < f_m(v)$.

Definition 3 (*Pareto optimal set*) For a given MOP $f(x)$, the Pareto optimal set is $P^* = \left\{x \in \Omega | \neg \exists\, x' \in \Omega, f(x') \preceq f(x)\right\}$.

Definition 4 (*Pareto optimal front*) For a given MOP $f(x)$ and its Pareto optimal set P^*, the Pareto front is $PF^* = \{f(x), x \in P^*\}$.

Definition 5 (*Ideal point*) The ideal point $Z^I = (Z_1^I, \ldots, Z_M^I)$ is the vector composed by the best objective values over the search space Ω. Analytically, the ideal objective vector is expressed by:

$$Z_m^I = Min_{x \in \Omega} f_m(x), m \in \{1, \ldots, M\} \tag{2}$$

Definition 6 (*Nadir point*) The nadir point $Z^N = (Z_1^N, \ldots, Z_M^N)$ is the vector composed by the worst objective values over the Pareto optimal set. Analytically, the nadir objective vector is expressed by:

$$Z_m^N = Max_{x \in P^*} f_m(x), m \in \{1, \ldots, M\} \tag{3}$$

Definition 7 (*ε-dominance*) A solution u is said to epsilon-dominate a solution v $(u \preceq_{\varepsilon+} v)$[1] if and only if $\forall m \in \{1, \ldots, M\} : u_m \leq v_m + \varepsilon$ for a given $\varepsilon > 0$, where u_m / v_m is the m-th objective value of solution u/v.

2 Resolution Methods

2.1 Aggregative Methods

Traditional multi-objective optimization methods aggregate the different objective functions into a single one. In order to generate a representative approximation of the whole Pareto front, the user must perform several runs with different parameter settings. Some representatives of this class of methods are the weighted sum method [2], the ε-constraint method [2], the goal programming [3], the reference point method [4], the reference direction method [5], and the light beam search method [6] which are briefly discussed in this subsection.

[1]We present the additive version of the ε-dominance. The multiplicative epsilon dominance is defined as follows: A solution u is said to epsilon-dominate a solution v $(u \preceq_\varepsilon v)$ if and only if $\forall m \in \{1, \ldots, M\} : u_m \leq v_m(1 + \varepsilon)$.

- **The weighted sum method**

This method converts the MOP to a single-objective optimization problem (SOP) by forming a linear aggregation of the objectives as follows:

$$\begin{cases} Min f(x) = w_1 f_1(x), w_2 f_2(x), \ldots, w_M f_M(x) \\ x \in \Omega \end{cases} \quad (4)$$

where w_m corresponds to the weighting coefficient of the m-th objective such that $\sum_{m=1}^{M} w_m = 1$ and $w_m \geq 0\ \forall m \in \{1, \ldots, M\}$. Solving (4) with different weighting coefficients sets yields a set of solutions. Under the condition that an exact optimization algorithm is used and all weighting coefficients are positive, it is easy to show that this method will only generate Pareto optimal solutions. Assuming that a feasible decision vector u minimizes f for a given weight combination and is not Pareto optimal, then there is a solution v which dominates u, i.e., $\forall m \in \{1, \ldots, M\}$ we have $f_m(v) \leq f_m(u)$ and $\exists m \in \{1, \ldots, M\}$ where $f_m(v) < f_m(u)$. Therefore, $f(v) < f(u)$, which is a contradiction to the assumption that $f(u)$ is minimum.

The main disadvantage of this technique is that it cannot generate all Pareto optimal solutions with non-convex trade-off surfaces. This is illustrated in Fig. 1a. For fixed weights w_1 and w_2, solution x is sought to minimize $y = w_1 f_1(x) + w_2 f_2(x)$. This equation can be formulated as $f_2(x) = -(w_1/w_2) f_1(x) + (y/w_2)$, which defines the line L (solid line in Fig. 1a) with a slope of $-(w_1/w_2)$ and an intercept of (y/w_2) in the objective space. Graphically, the optimization process corresponds to moving this solid line downwards until no feasible objective vector is above it and at least one feasible objective vector (here A and D) is on it. However, the points B and C will never minimize y. In fact, if the slope is increased (upper dashed line), D achieves a lesser value of y than B and C. Besides, if the slope is decreased (lower dashed line), A has a lesser y value than B and C.

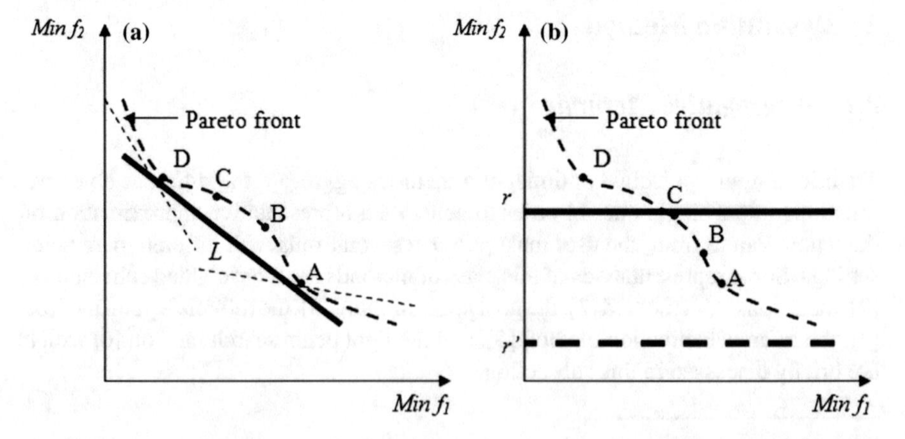

Fig. 1 Graphical interpretation of: **a** the weighted sum method and **b** the ε-constraint method (inspired by [7])

- **The ε-constraint method**

This method converts the MOP into a SOP by optimizing individually a selected objective while keeping the remaining $(M-1)$ objectives' values less than or equal to some user-specified thresholds as follows:

$$\begin{cases} Min f(x) = f_h(x) & h \in \{1, \ldots, M\} \\ f_m(x) \leq \varepsilon_m & m \in \{1, \ldots, M\}, m \neq h; \\ x \in \Omega \end{cases} \tag{5}$$

The upper bounds ε_m are the parameters to be varied in each run in order to obtain multiple Pareto optimal solutions. As depicted in Fig. 1b, the ε-constraint method is able to find solutions associated with non-convex parts of the Pareto front. Setting $h = 1$ and $\varepsilon_2 = r$ (solid line in Fig. 1b) makes solution D infeasible while solution C minimizes f_1. Figure 1b also shows a problem with this technique. In fact, if the lower bounds are not chosen appropriately ($\varepsilon_2 = r'$), the obtained feasible set may be empty, i.e., there is no solution to the obtained SOP. In order to avoid this problem, a suitable range of values for the ε_m quantities has to be known beforehand.

- **The goal-programming method**

For each objective function, the user provides a goal G_i to be achieved. The goal programming method transforms the MOP into a SOP by minimizing individually the weighted sum of deviations from goals as follows:

$$\begin{cases} Min f(x) = \sum_{m=1}^{M} w_m |f_m(x) - G_m| \\ x \in \Omega \end{cases} \tag{6}$$

where w_m corresponds to the weighting coefficient of the m-th objective such that $\sum_{m=1}^{M} w_m = 1$ and $w_m \geq 0 \; \forall m \in \{1, \ldots, M\}$.

As discussed by [8], if the optimal objective function value of the goal programming method equals zero, then some caution is in order since the obtained solution may not be Pareto optimal. In fact, if all settled goals are feasible, then the value zero for all the deviational variables gives the minimal value (zero) for the goal programming objective function. Hence, the solution is equal to the reference point (the vector composed with all user-specified goals) and normally there exist many feasible solutions that are non Pareto optimal. If the solutions are intended to be Pareto optimal independently of the selection of goals, then if the goals are feasible, the function f is to be maximized; else if the goals are infeasible the function f is to be minimized.

- **The reference point method**

The classical Reference Point Method (RPM) was proposed by [4]. A reference point g for a particular MOP consists of an aspiration level vector. Aspiration levels represent the DM's desired values for each objective. This method projects the reference point onto the Pareto optimal region via the minimization of an Achievement Scalarizing Function (ASF). Among the most commonly known forms of an ASF is the following:

$$Min\ s(f(x), g) = \underset{m=1,\dots,M}{Max}\ [w_m(f_m(x) - g_m)] \tag{7}$$

where g_m is the m-th component of the reference point and w_m is the weight associated with the m-th objective.

As shown in Fig. 2, the reference point could be feasible belonging to the Pareto front (a), feasible not belonging to the Pareto front (b) or infeasible (c). For a chosen reference point, the RPM tries to find the closest Pareto optimal solution. The main drawback of this method is that it provides only one solution in a single run. Hence, if the DM is dissatisfied with the obtained solution and/or he/she would like to obtain a small sample of Pareto optimal solutions near each reference point then he/she must perform several runs of the algorithm. It should be noted that the DM could obtain a sample of near reference point solutions by perturbing the reference point and/or the weights and performing several runs of this method. Besides, in order to make this method interactive, Wierzbicki [4] suggested a procedure to update the reference point automatically which facilitates the DM's task. When using the reference point approach in practice, the DM is asked to supply a reference point and a weight vector at a time. The reference point guides the search towards the desired region while the weight vector provides more detailed information about which Pareto optimal point to converge to.

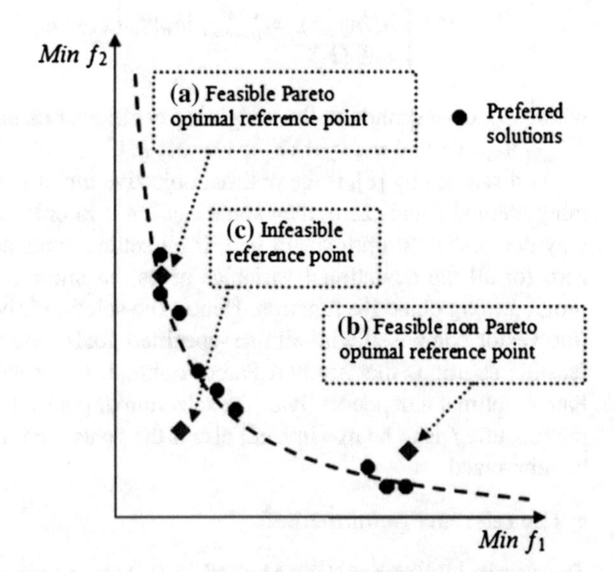

Fig. 2 The reference point method

- **The reference direction method**

Korhonen and Laakso [5] suggested a reference direction-based approach for multi-criterion optimization using the principle of solving ASFs repeatedly. This method is described as follows:

- **Step 1**: Choose an initial arbitrary point q^0 in the objective space and let $K \leftarrow 1$;
- **Step 2**: Specify another vector g^k and determine the reference direction $d^k = g^k - q^{k-1}$;
- **Step 3**: Determine a set Q^k of efficient solutions q which solves the following ASF:

$$\begin{cases} Min(f(x), r, w) = \underset{w_m > 0, m=1,\ldots,M}{Max} [(f_m(x) - r_m(t))/w_m] \\ r(t) = q^{k-1} + td^k \end{cases} \quad (8)$$

where t is an integer parameter increased from zero to infinity, w is a weighting vector, and $r_m(t)$ is the m-th component of $r(t)$;

- **Step 4**: Find the most preferred solution q^k in Q^k using a particular utility function or by other mean;
- **Step 5**: If $q^{k-1} \neq q^k$, set $k \leftarrow k + 1$ and go to **Step 2**. Otherwise, check for optimality conditions (Kuhn-Tucker conditions [8] or other optimality conditions [5]) of the solution q^k. If q^k is optimal then terminate the optimization run. Otherwise, increment k, determine a new reference direction and go to **Step 3**.

Figure 3 shows a sketch of **Step 3** of the above optimization procedure. For each point (say point C) marked on the reference direction (from q^0 towards g^1), a Pareto optimal solution (point A) is found by solving the ASF given in Eq. (8). **Step 3** of the above procedure involves multiple application of a single-objective optimization for different values of t, thereby finding a range of efficient solutions (A till E). The idea of finding an efficient solution corresponding to a point on a reference direction is similar to the reference point approach of Wierzbicki [4]. Although the original study of the reference direction approach and subsequent studies of Korhonen and his co-authors [10, 11] concentrated on parametric solutions for multiple points on the reference direction, the principle can be used by forming multiple ASFs and solving them by a single-objective optimizer independently. An analytical hierarchy process was also used to determine the reference direction [12]. Interestingly, the reference direction approach corresponds to the process of projecting the reference direction on the Pareto optimal frontier.

- **The light beam search method**

The Light Beam Search (LBS), as described in [13], combines the reference point idea and tools of Multi-Attribute Decision Analysis (MADA). It enables an interactive analysis of MOPs thanks to the presentation of samples of a large set of non-dominated points to the DM in each iteration. An aspiration point and a reservation one should be supplied by the DM. These two points define the direction of the

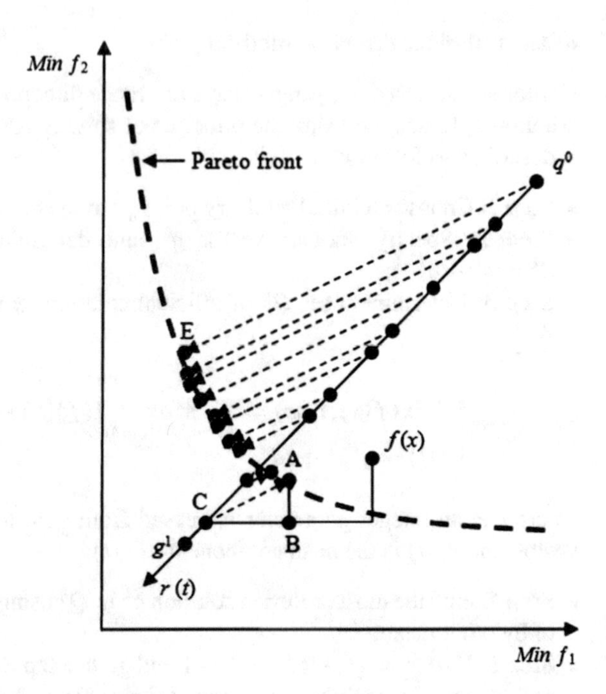

Fig. 3 The reference direction method [9]

search in a particular iteration. If these two points are not suggested, the ideal point and the nadir point (or a worse point than the nadir one) can be assumed as aspiration and reservation points respectively. Initially a non-dominated middle point is determined by projecting the aspiration point on to the non-dominated front by using an augmented version of Wierzbickis ASF. Thereafter, a local preference model in the form of an outranking relation *S* is used to obtain neighboring solutions of the current non-dominated point, or the middle point. It is said that a outranks *b* (or *a S b*), if *a* is considered to be at least as good as *b*. To define an outranking relation, the DM has to specify three preference thresholds for each objective: (1) indifference threshold, (2) preference threshold, and (3) veto threshold. In the LBS procedure, they are considered to provide only local information, thus they are assumed to be constants. The extreme points or characteristic neighbors are found one for each objective by considering the maximum allowed improvement in a particular objective in relation to the middle point. The DM can control the search by either modifying the aspiration and/or reservation points, or by shifting the middle point to selected better point from its neighborhood or by modifying the preference threshold values. Figure 4 illustrates the LBS method mechanism. The LBS procedure is as follows:

- **Step 1**: Ask the DM to specify starting aspiration and reservation points;
- **Step 2**: Compute the starting middle point on the Pareto optimal front;
- **Step 3**: Ask DM to specify the local preferential information used to build an outranking relation;
- **Step 4**: Present the middle point to the DM;

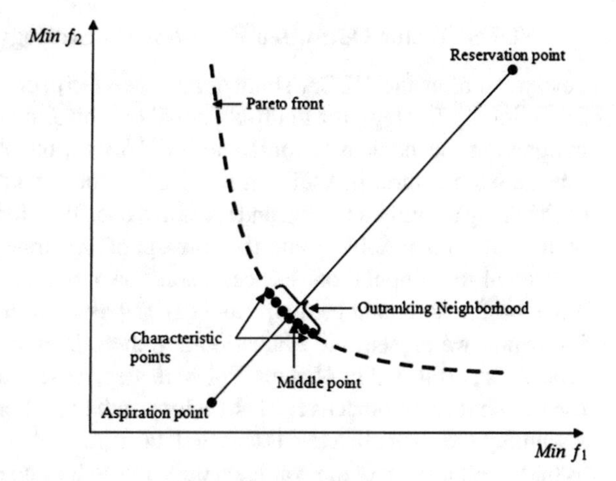

Fig. 4 The light beam search method

- **Step 5**: Calculate the characteristic neighbors of the middle point and present them to the DM;
- **Step 6**: If DM is satisfied, terminate the procedure; else ask the DM to: (1) choose one of the neighboring points to be the new middle point, (2) update the preferential information, or (3) define a new aspiration point and/or a new reservation point. The algorithm proceeds by moving to **Step 5** for the case (1) and to **Step 4** otherwise.

2.2 Evolutionary Methods

- **Non Pareto-based evolutionary methods**
  - **VEGA: Vector Evaluated Genetic Algorithm**

Schaffer [13] proposed one of the first alternatives to adapt EAs to handle MOPs called VEGA. The basic idea is to divide the population into *M* subpopulations of equal sizes. Then, in each one of them, the selection operates by taking into account only the unique corresponding objective. Once the selection mechanism was performed, the population is mixed to apply the rest of the evolutionary operators. All this process is repeated in each generation. An evident VEGA problem is that it does not promote the survival of good trade-off solutions, but it prefers the best solutions of each objective separately. This problem is known as speciation (by its analogy in genetics). This problem was identified and attacked by Schaffer, using mating restrictions (i.e., not allowing recombination between individuals of the same subpopulation) as well as other heuristic rules applied during the selection mechanism. In another work [14], it was also demonstrated that, if proportional selection is used, VEGAs scheme is equivalent to a linear combination of objective functions which means that it has limitations regarding non-convex Pareto fronts.

– **VOES: Vector Optimized Evolutionary Strategy**

Few years after the VEGA studies, Kursawe [15] proposed the Vector Optimized Evolutionary Strategy for multi-objective optimization (VOES). The VOES fitness assignment mechanism is similar to VEGA one, but Kursawe used other genetic aspects from nature. In VOES, a solution is represented by a diploid chromosome, each having a dominant string and recessive one. Two different solution vectors (each with a decision variable x and the corresponding strategy vector σ) are used as an individual in a population. Hence, a solution x is evaluated by calculating: (1) f^d based on the dominant genotype and (2) f^r based on the recessive genotype. In the following, we present the evaluation and the selection mechanisms. The selection process is performed in M steps. For each step, a user-supplied probability vector is used to choose an objective. This vector can be fixed or varied across generations. Assuming the m-th objective is selected, the fitness of certain solution x is computed as the weighted sum of the dominant objective value and the recessive one as follows:

$$f(x) = \frac{2}{3} f^d_m(x) + \frac{1}{3} f^r_m(x) \tag{9}$$

For each selection step, the population is sorted based on each objective function and $(\frac{M-1}{M})$th portion of the population is selected as parents. This procedure is repeated M times, every time using the survived population from the previous sorting. Thus, the relation between the number of parents μ and the number of children λ can be expressed as follows:

$$\mu = \left(\frac{M-1}{M}\right)^M \lambda \tag{10}$$

For example, for the bi-objective case, we obtain $\mu = 0.25\lambda$. All new μ solutions are copied into an external archive which stores the non-dominated individuals found since the beginning of the simulation run. After adding such solutions to this archive, a non-domination check is performed and only new non-dominated solutions are retained. If the size of the external archive exceeds the archive size, a niching mechanism is used to eliminate crowded solutions in order to promote diversity.

VOES uses non-domination check to ensure convergence and niching to encourage diversity. These features are essential to design a good MOEA. Unfortunately, Kursawe assessed the performance of his algorithm on a single test problem and no further experimental assessments were pursued since Kursawe's original study.

– **WBGA: Weight-Based Genetic Algorithm**

WBGA, also called HLGA (Hajela and Lin Genetic Algorithm), was introduced by [16]. For each objective function, a weighting coefficient is assigned. Unlike the classical weighted sum method, each individual from the population has its own weighting coefficient vector which is coded in its string concatenated to its decision variables. This fact makes the WBGA able to find multiple non-dominated solutions in a single run. The key issue in this algorithm is how to maintain the diversity of weighting coefficients among the population individuals. Tow approaches were suggested for this sake. In the first approach, a niching mechanism is used on the substring representing the weight coefficient vector. In the second approach, carefully chosen subpopulations are evaluated for different pre-defined weight vectors in a similar way to VEGA. Unfortunately, WBGA is a weight-based approach; hence it fails in finding Pareto optimal solutions residing in the non-convex parts of the front.

- **Pareto-based evolutionary methods**
  - **Non elitist methods**

    MOGA: Multi-objective Genetic Algorithm

MOGA [17] is the first MOEA which explicitly used Pareto-based ranking and niching techniques together to encourage the search towards the true Pareto front while maintaining diversity in the population. In fact, each individual is assigned a rank which is expressed as a function of the number of individuals dominating it. Assuming $Ndom^t$ to be the number of solutions dominating a certain solution x at a generation t, the rank at t of x is given by:

$$rank^t(x) = 1 + Ndom^t \quad (11)$$

With such ranking mechanism, non-dominated solutions have a rank of 1 (cf. Fig. 5). The fitness assignment method used in MOGA takes into account the rank of the population member and the average fitness value of the population. The process for computing the fitness values is as follows. Firstly, the population is sorted by rank. Then, a fitness value is assigned to each individual based on an interpolation of the

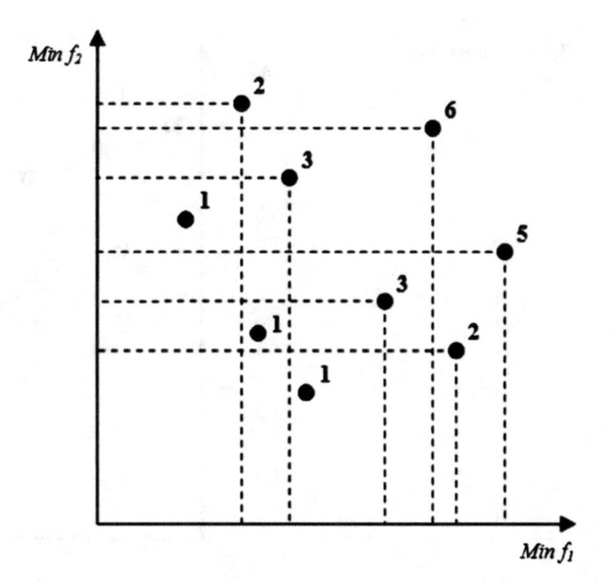

Fig. 5 MOGA ranking process

best rank to the worst rank according to some specified function. Finally, individuals assigned the same rank receive an averaged fitness value. This ensures that all population members of the same rank are sampled with an identical frequency. This information is used to maintain constant global population fitness with an appropriate amount of selective pressure. Additionally, MOGA implements the concept of fitness sharing (also referred to as crowding or niching) and uses a σ_{share} parameter called the niche radius which must be carefully specified. The niching mechanism is applied in the objective space in order to obtain a uniform distribution of the Pareto front approximation. Figure 6 illustrates the fitness sharing mechanism. In fact, solutions residing inside the niching radius are penalized in their fitness values. Although in MOGA fitness assignment is explicitly based on Pareto dominance, solutions having the same rank may not have the same assigned fitness. This may cause an unwanted bias towards a certain zone of the search space. Particularly, MOGA may be sensitive to the geometry of the Pareto front in addition to the density of solutions over the search space. Besides, the fitness sharing mechanism favors solutions with poor ranks over solutions with higher ranks if these latter are more crowded, thereby worsening the converging.

NPGA: Niched Pareto Genetic Algorithm

Horn and Nafpliotis [18] proposed NPGA which differs from the previously discussed MOEAs in the selection operator. This algorithm uses the binary tournament selection instead of proportionate selection methods used in VEGA and MOGA. During the tournament selection, two solutions x and y are picked randomly from the parent population P. Then, these two solutions are compared based on Pareto dominance to each individual of a randomly selected subpopulation T of size $tdom$ (where $tdom \ll |P|$). If one of the two solutions is non-dominated with respect to

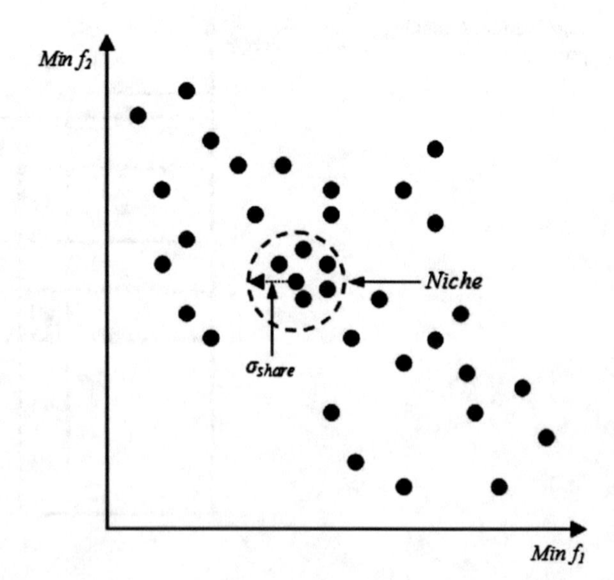

Fig. 6 Fitness sharing strategy

all the subpopulation individuals and the other one is dominated by at least one individual, the non-dominated solution is retained. In the cases where neither or both members are dominated by the subpopulation members, a niching mechanism is implemented to select the least crowded solution among x and y.

NPGA is found to be sensitive to the σ_{share} value in addition to the $tdom$ one. The numerical results reported in [18] suggest that $tdom$ should be an order of magnitude smaller than the population size. On one hand, if $tdom$ is too small, the non-domination check would be so noisy which may not emphasize non-dominated solutions sufficiently. On the other hand, if is too large, non-dominated solutions will be well-emphasized but the computational complexity will increase. Additionally, $tdom$ depends on the number of objectives to optimize.

NSGA: Non-dominated Sorting Genetic Algorithm

NSGA [19] is based on the non-dominated sorting strategy (cf. Fig. 7). This strategy classifies the population members into several fronts. The non-dominated sorting algorithm begins by identifying the non-dominated individuals from all population members. These individuals have the rank of one and are assigned a large dummy fitness value. After that, the first front members are discarded temporary from the population and the non-dominated individuals from the truncated population are identified and assigned the rank of 2 (eventually assigned a dummy fitness value smaller than the one of the first front). This process continues until classifying all population members. The diversity maintenance is achieved in NSGA by applying the fitness sharing front-wise in the decision space (instead of the objective space) in order to degrade the fitness values based on a user-defined niche radius value σ_{share}. The sharing in each front is achieved by calculating a sharing function value between two individuals i and j in the same front as follows:

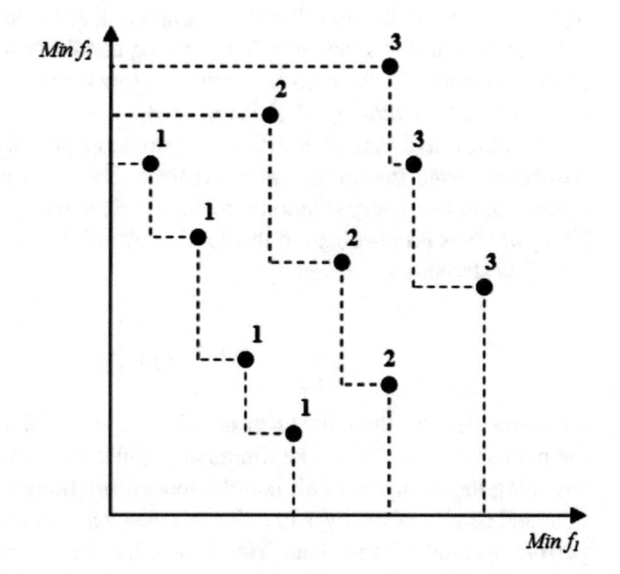

Fig. 7 Non-dominated sorting strategy

$$\begin{cases} Sh_{d_{ij}} = 1 - (\frac{d_{ij}}{\sigma_{share}})^2 & \text{if } d_{ij} < \sigma_{share} \\ 0 & \text{otherwise} \end{cases} \quad (12)$$

where d_{ij} is the Euclidean distance separating i and j. After that, a parameter niche count is calculated by adding the above sharing function values for all individuals in the current front. Finally, the shared fitness value of each individual is computed by dividing its dummy fitness value by its niche count. The best individuals are always preferred over other solutions, thereby favoring the generation of new individuals near the non-dominated solutions. The fitness sharing mechanism helps the algorithm to distribute the non-dominated solutions along the Pareto front. However, the high sensitivity to the σ_{share} parameter yields to a less efficient performance of NSGA.

– **Elitist methods**

Elitism means that elite individuals cannot be excluded from the archive gene pool of the population in favour of worse individuals [20]. In the following, we review the most representative elitist MOEAs [21].

SPEA/SPEA2: Strength Pareto Evolutionary Algorithm

Zitzler and Thiele [22] proposed the strength Pareto approach which uses two populations: (1) a main population P and (2) an archive population A which contains the non-dominated individuals found so far during the evolutionary process. Initially, the population P is generated randomly and the archive A is empty. Then, A is filled with non-dominated members from P. After that, solutions from A which are dominated by any other member from A are deleted. Besides, if the number of externally stored non-dominated solutions exceeds the archive size $|A|$, then A is pruned by means of a clustering procedure which will be discussed next. Once all population and archive members are each assigned a fitness value, binary tournament selection with replacement is applied to fulfill the mating pool. After applying genetic operators, a new population P is generated. If a stopping condition is met then the evolutionary process is stopped, else non-dominated vectors from P are copied to the archive A as usual and the overall process is repeated.

The fitness assignment in SPEA is a two-stage process. First, the non-dominated individuals from the archive A are ranked. Then, the population P members are evaluated. In fact, every solution i from the archive A is assigned a strength value $s_i \in [0, 1[$ which is proportional to the number of individuals in P which are dominated by i. The strength s_i is given by:

$$s_i = \frac{nd}{|P| + 1} \quad (13)$$

where nd denotes the number of individuals in P that are covered by i and $|P|$ is the main population size. The fitness of population individual $j \in P$ is obtained by summing the strengths of all non-dominated solutions $i \in A$ that dominates j. The obtained sum is raised by 1 in order to guarantee that archive members have better performance than P members. This fitness is to be minimized and is given by:

$$f_j = 1 + \sum_{i,i \preceq j} s_i \tag{14}$$

The clustering mechanism is applied to reduce the size of the archive while keeping its characteristics. The general idea is to partition the archive into C groupings (clusters), where $C < |A|$ and all individuals of the same grouping have the same characteristics. The clustering procedure begins by making each element of the initial non-dominated archive a cluster. Following this, two clusters are chosen via a distance measurement to be combined into one cluster. The distance is calculated as average Euclidean distance between pairs of individuals across the clusters. At the completion of the clustering process, the new non-dominated archive consists of the centroid members of each cluster. The authors show favorable results compared to other MOEAs.

In another study [23] have identified three weaknesses for SPEA. Firstly, for the fitness assignment strategy, individuals that are dominated by the same archive members have identical fitness values. Hence, in the case when the archive contains only a single individual, all population members have the same rank independently of whether they dominate each other or not. Consequently, the selection pressure is decreased substantially and in this particular case SPEA behaves like a random search algorithm. Secondly, for the density estimation, if many individuals of the current generation are Pareto equivalent, none or very little information can be obtained on the basis of the partial order defined by the dominance relation. In this situation, which is very likely to occur when the number of objectives exceeds two, density information has to be used in order to guide the search more effectively. Clustering makes use of this information, but only with regard to the archive and not to the main population. Thirdly, for the archive truncation strategy, although the clustering mechanism used in SPEA is able to reduce the non-dominated set without destroying its characteristics, it may lose extreme (outer) solutions. However, these solutions should be kept in the archive in order to obtain a good spread of non-dominated solutions. In response to the above mentioned SPEA weaknesses, Zitzler et al. [23] have proposed an improved version of SPEA, called SPEA2. In contrast to SPEA, SPEA2 uses a fine-grained fitness assignment strategy which incorporates density information. Furthermore, the archive size is fixed, i.e., whenever the number of non-dominated individuals is less than the predefined archive size, the archive is filled up by dominated individuals; with SPEA, the archive size may vary over time. In addition, the clustering technique, which is invoked when the non-dominated front exceeds the archive limit, has been replaced by an alternative truncation method which has similar features but preserves boundary solutions. Finally, another difference to SPEA is that in SPEA2 only members of the archive participate in the mating selection process.

The SPEA2 fitness assignment for a certain solution i takes into account the number of individuals dominating i in addition to the number of individuals dominated by i. Each solution i from the population P and the archive A is assigned a strength value s_i representing the number of individuals dominated by i:

$$s_i = |j | j \in P \cup A \wedge i \preceq j| \tag{15}$$

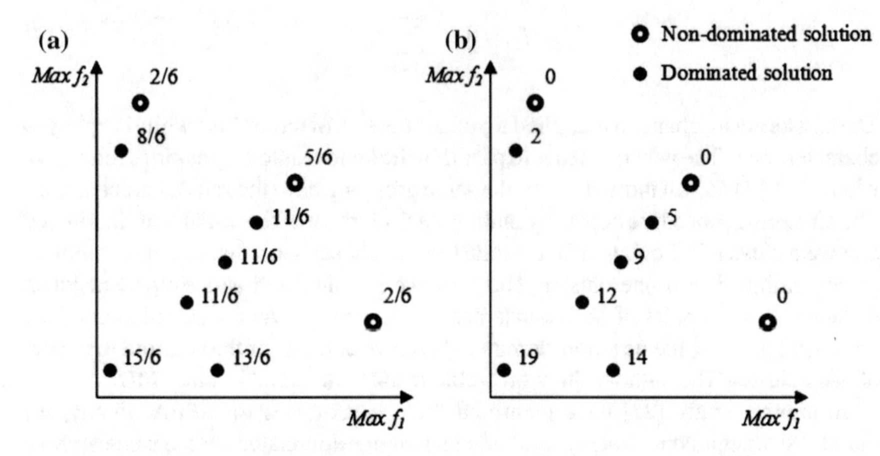

Fig. 8 Comparison of fitness assignment mechanisms: **a** SPEA versus **b** SPEA2 (from [23])

After that, the raw fitness R_i is computed as flows:

$$R_i = \sum_{j \in P+A, j \preceq i} s_j \tag{16}$$

This raw fitness is determined by the strengths of its dominators in both archive and population, as opposed to SPEA where only archive members are considered in this context. It is important to note that fitness is to be minimized here, i.e., $R_i = 0$ corresponds to a non-dominated individual, while a high R_i value means that i is dominated by many individuals (which in turn dominate many individuals). This scheme is illustrated in Fig. 8b.

The raw fitness assignment strategy supplies a sort of niching based on the Pareto dominance concept. However, this strategy becomes inefficient when most individuals are non-dominated with each other. For this reason, additional density information is incorporated to discriminate between individuals having identical raw fitness values. The density estimation technique used in SPEA2 is an adaptation of the k-th nearest neighbor method where the density at any point is a decreasing function of the distance to the k-th nearest points. The density estimate corresponds to the inverse of the distance to the k-th nearest neighbor. In fact, for each individual i, the distances in objective space to all individuals j from $P \cup A$ are computed then stored in a list in an increasing order. After that, the k-th nearest neighbor gives the sought distance denoted by σ_i^k. The k parameter value is usually set to $\sqrt{|P| + |A|}$. The density D_i of solution i is:

$$D_i = \frac{1}{\sigma_i^k + 2} \tag{17}$$

In the denominator, two is added to ensure that its value is greater than zero and that $D_i < 1$ Finally, the fitness of a certain solution i is obtained by summing the raw fitness and the density information as follows:

$$F_i = R_i + D_i \tag{18}$$

The SPEA2 environmental selection mechanism differs from SPEA one by preserving the boundary solutions and by the fact that the number of stored external solutions is constant over time.

NSGA-II: Non-dominated sorting Genetic Algorithm II

NSGA-II is the improved version of NSGA [24, 25]. NSGA-II is one of the most cited MOEAs. The most prominent features of NSGA-II are its low computational complexity, elitist approach and a method for diversity that does not need additional parameters. The general principle of NSGA-II is as follows. The NSGA-II algorithm begins by creating an offspring population Q_0 by applying genetic operators to a randomly generated parent population P_0. From the first generation award, the basic iteration of NSGA-II is different. First, the two populations P_t and Q_t are combined to form a population R_t of size $2N$ ($|P_t| = |q_t| = N$). Second, a non-dominated sorting is performed to classify the entire population R_t. Once, the non-dominated sorting is over, the population R_t becomes subdivided in several categories in the same manner of NSGA. After that, the new parent population P_{t+1} is filled with individuals of the best non-dominated fronts, one at a time. Since the overall population size is $2N$, not all fronts may be accommodated in N slots available in the new population P_{t+1}. When the last allowed front is being considered, it may contain more solutions then the remaining available slots in P_{t+1}. Instead of discarding arbitrary some elements from the last front, NSGA-II uses a niching strategy to choose individuals from the last front which reside in the least crowded regions in this front. In fact, for each ranking level, a crowding distance is estimated by calculating the sum of the Euclidean distances between the two neighboring solutions from either side of the solution along each of the objectives as demonstrated by Fig. 9. In order to preserve boundary solutions, these latter are each assigned an infinite crowding distance. The crowding distance assignment procedure can be summarized by the three following steps:

- **Step 1**: For each solution i from the considered front F, initialize its crowding distance CD_i to zero: $CD_i \leftarrow 0$;
- **Step 2**: For each objective function, sort the front members in a decreasing order of f_m, and find the sorted indices vector: $I^m = sort(f_m, >)$;
- **Step 3**: For $m = 1, \ldots, M$, assign an infinite crowing distance to extreme solutions ($CD_{I_1} = CD_{I_F} = \infty$) and for the other solutions $j = 2, \ldots, |F| - 1$, assign:

$$CD_{I_j^m} = CD_{I_j^m} + \frac{f_m^{I_{j+1}^m} - f_m^{I_{j-1}^m}}{f_m^{max} - f_m^{min}} \tag{19}$$

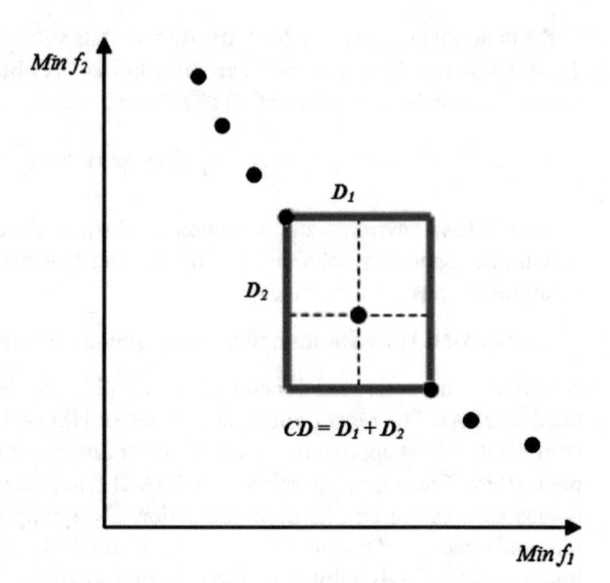

Fig. 9 Crowding distance

where I_j^m corresponds to the index of the j-th member in the list sorted based on the m-th objective function. NSGA-II is demonstrated to be one of the most competitive MOEAs through the specialized literature. The main weakness of NSGA-II was reported in [1]. In fact, when the cardinality of the first front from the combined population R_t exceeds the population size $|P|$, some closely packed Pareto optimal solutions may give their places to some non-dominated yet non Pareto optimal solutions since the replacement becomes based only on the crowding distance criterion.

PAES/PESA: Pareto Archived Evolutionary Strategy/Pareto Envelope-based Selection Algorithm

Knowles and Corne [26, 27] proposed a $(1+1)$-Evolutionary Strategy ($(1+1)$-ES), named PAES, to approximate the whole Pareto front. This work was motivated by the success of (1+1)-ES in resolving mono-objective problems. For this reason the authors have adapted such search method for the multi-objective case. PAES begins by producing a child c_0 from a randomly generated parent p_0. In each generation t, non-dominated solutions found are stored in an archive with a pre-specified size. The two individuals p_t and c_t are firstly compared. If one solution dominates the other, the dominated individual is discarded and the dominant one is retained as parent for the next generation. In the case where p_t and c_t are non-dominated, the new candidate solution is compared with a reference population of previously archived non-dominated solutions, i.e., archive members. If comparison to the population in the archive fails to favor one solution over the other, the tie is split to favor the solution which resides in the least crowded region of the search space. The archive has a user-specified maximum size which reflects the desired number of final solutions. Each child c_t which is not dominated by its parent p_t is compared with each member of the archive. Candidates which dominate the archive members are always accepted (as parents) and archived. Candidates which are dominated by the archive members are always rejected, while those which are non-dominated are accepted and/or archived based on the degree of crowding in their grid location. The major feature of PAES is its strategy for promoting diversity in the approximation set. PAES uses an adaptive hyper-gridding system in the objective space to divide it into d non-overlapping hyper-boxes. The belonging of a certain solution to a certain region in the hyper-box is determined by the objectives' values which define the solution's coordinates. In the case where an offspring solution is non-dominated with respect to the archive members, a crowding measure based on the number of solutions residing in a certain hyper-box is applied to determine whether the offspring solution is accepted or not.

The major advantage of this diversity maintenance technique is that it does not require any additional parameters such as the niche size parameter σ_{share}. However, the main crux of PAES is the sensitivity of the performance of such algorithm to the d parameter of the hyper-gridding system (cf. Fig. 10).

The same authors [28] have proposed PESA which is a modified version of PAES. PESA has the same archiving and diversity preserving mechanisms of PAES. In PESA, like SPEA2, only archive members participate in genetic operations. PESA begins by randomly generating a small internal population *IP*. PESA uses also a large external population *EP* which is initially empty. After that, the archive *EP* is updated with elite solutions in the same manner as PAES. If the stopping criterion is met then the algorithm returns *EP*, else *IP* is fulfilled with new individuals by the following operations. With probability p_c, two parents are selected from *EP*. A single child is subsequently created by crossover. This child is then mutated. With probability

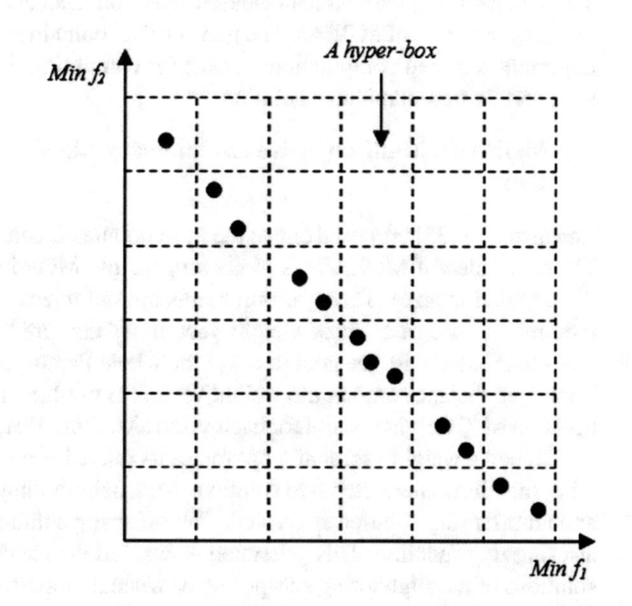

Fig. 10 PAES hyper-gridding system with $d = 6$

$(1 - p_c)$, a selected parent from *EP* is mutated. After that, the archive *EP* is updated and the overall process is repeated.

As PAES, PESA necessitates the tuning of the archive size and the d parameter of the gridding system. We note that the number of hyper-boxes changes exponentially with the modification of d value which influences the final population distribution. An improved version of PESA, called PESA-II, was proposed by Corne et al. [29] where selection is region-based and the subject of selection is now a hyper-box not only an individual (i.e., first selecting a hyper-box, then an individual is chosen from the selected hyper-box). The motivation behind PESA is to reduce the computational cost of Pareto ranking.

IBEA: Indicator-Based Evolutionary Algorithm

Zitzler and Künzli [30] proposed a MOEA where selection is based on solution contribution to a certain quality indicator. Indicator-based MOEAs can, therefore, be seen as a third generation of MOEAs. IBEA begins by randomly generating a population P. After that, for each solution i from P, the algorithm computes the fitness of i corresponding to the loss in quality if i is removed from the population P. The solution with the lowest fitness is removed from the population and then the population members' fitness values are recomputed since the population is truncated. This selection strategy is used in creating the mating pool and in environmental selection. The main crux of IBEA is its sensitivity to the κ parameter which is used to scale the fitness function values since the algorithm performance largely depends on this parameter which is reported to depend of the considered MOP. Another indicator-based selection algorithm is the S Metric Selection-based Evolutionary Multi-objective Algorithm (SMS-EMOA) [31] which combines non-dominated sorting with indicator-based selection mechanism. IBEAs can be seen as the last generation of MOEAs. The main critical point in this type of algorithms is the important required computational effort for computing the quality indicator values for a certain non-dominated solution set [32].

MOEA/D: Multi-objective Evolutionary Algorithm Based on Decomposition

Zhang and Li [33] proposed one of the most popular decomposition-based algorithm. The basic idea of MOEA/D is to decompose the MOP into N sub-problems (N is the population size). These sub-problems are optimized simultaneously. MOEA/D requires the use of a set of weight vectors λ_j that are generated so that they are well-distributed with the goal to cover the whole Pareto front. In their approach, the Euclidean distance among the weight vectors is used in order to determine a neighborhood of T weight vectors for each vector λ_j. After that, each population member is assigned a weight vector and optimizes its related sub-problem based on a scalarizing function. Thereafter, two solutions from neighboring weight vectors are mated and an offspring solution is created. The offspring solution is then evaluated using a scalarizing function. This generated new solution can also replace several current solutions of its neighboring sub-problems when it outperforms them. Three versions of scalarizing functions are adopted for MOEA/D: (1) weighted sum approach, (2) weighted Tchebycheff approach, and (3) boundary intersection approach. The diversity in MOEA/D is managed based on similarity between individuals weight vectors, i.e., based on a corresponding neighborhood of solutions. MOEA/D is good in finding a small number of uniformly distributed Pareto solutions at low computational cost. Moreover, MOEA/D has demonstrated very interesting results on several problems with a high number of objectives. However, its main shortcoming is the degradation of diversity and solution distribution when tackling badly-scaled problems (i.e., problems where the objective functions do not have the same scale).

3 Performance Assessment of MOEAs

3.1 Test Functions

Several test functions are proposed to challenge MOEA capabilities in approximating the Pareto front. The most cited test function suites are: (1) the bi-objective ZDT (Zitzler-Deb-Thiele) suite [34] and (2) the scalable DTLZ (Deb-Thiele-Laumans-Zitzler) suite [35] where the Pareto optimal front can be determined analytically. Such test functions encapsulate several characteristics such as non-convexity, multimodality, non-uniformity of the search space, and discontinuity which cause difficulties to a MOEA. These test functions do not reflect necessarily the main features of real world MOPs. It is true that some of these functions contain important characteristics that make them particularly difficult to solve. Thus, if a MOEA can resolve such test functions, it should also be able to tackle real world MOPs; following a well-defined adaptation step. Tables 1 and 2 present the ZDT and DTLZ test functions characteristics respectively. We notice that for DTLZ test problems, the parameters can be modified in order to increase or decrease the problems difficulties (e.g., modifying the number of local optimal Pareto fronts).

Table 1 Bi-objective ZDT test problems' characteristics

Name	Features
ZDT1	The Pareto front is convex
ZDT2	The Pareto front is concave
ZDT3	The Pareto front is formed by several disjoint convex parts
ZDT4	There are 21^9 local fronts
ZDT5	The Pareto front is convex. ZDT5 is a discrete problem with a deceptive landscape
ZDT6	The Pareto front is concave. This problem is characterized by the non-uniformity not only of the search space but also of the solution distribution along the Pareto front

Table 2 Scalable DTLZ test problems' characteristics

Name	Features
DTLZ1	The Pareto front is linear (Hyper-plane). There are $(11^k - 1)$ local optimal fronts where k is a user-specified parameter
DTLZ2	For $M > 3$, the Pareto optimal solutions lie inside the first quadrant of the unit sphere in a three-objective plot with f_M as one of the axes
DTLZ3	There are $(3^K - 1)$ local fronts that are parallel to the global Pareto front where k is a user-specified parameter
DTLZ4	The Pareto optimal solutions are non-uniformly distributed along the Pareto front
DTLZ5	The front is a curve and the Pareto optimal solutions are non-uniformly distributed along the Pareto front
DTLZ6	The front is a curve and the solution density gets thinner towards the Pareto front
DTLZ7	The Pareto front is formed by 2^{M-1} disjoint regions in the objective space
DTLZ8	The Pareto front is a combination of a straight line and a hyper-plane. The straight line is the intersection of the first $(M - 1)$ constraints with $f_1 = f_2 = \cdots = f_{M-1}$ and the hyper-plane is represented by another constraint g_M
DTLZ9	The Pareto front is a curve with $f_1 = f_2 = \cdots = f_{M-1}$. The solution density gets thinner towards the Pareto front

3.2 *Performance Indicators*

When evaluating the performance of a MOEA, there are two main goals to pursue: (1) closeness of the provided non-dominated solution set to the Pareto optimal front and (2) diversity of the obtained solution set (with a good distribution) along the Pareto optimal front. Several performance measures are proposed in the EMO literature to evaluate one or both of these goals [36]. Table 3 presents a classification of selected representative performance measures. The classification criteria are the following:

- *unary* which indicates if it is a unary performance indicator (i.e., performance measure which assigns a single value to each non-dominated solution set);
- *binary* which indicates if it is a binary performance indicator (i.e., performance measure which assigns a single value to a pair of non-dominated solution sets);
- *convergence* which indicates that the performance indicator assigns a single value corresponding to the convergence of the non-dominated solution set;
- *diversity* which indicates that the performance indicator assigns a single value corresponding to the diversity of the non-dominated solution set;
- *reqPFtrue* which indicates if the performance measure requires the true Pareto optimal front PF_{true} to assign a single value to the non-dominated solution set;
- *bestvalue* which indicates the best value that can be obtained from the performance indicator;
- *Pareto compliant* which indicates whether the performance measure is Pareto dominance compliant. Before defining the notion of Pareto dominance compliance,

Table 3 Main features of performance indicators

Performance indicators	Unary	Binary	Convergence	Diversity	reqPFtrue	Best value	Pareto compliant
EC	X		X		X	0	X
SC		X	X			1	X
$I_{\varepsilon+}$		X	X			-	X
GD	X		X		X	0	
IGD	X		X		X	0	
Δ	X			X		0	
HV	X		X	X		1	X
S	X			X		0	
R2	X		X	X		0	X
χ^2-like deviation	X			X	X	0	

we give the definitions of compatibility and completeness. The definitions are derived from the study of Zitzler et al. [36]:

Definition 8 (*Compatibility and Completeness*) Assuming W and Z two approximation sets, a quality indicator $I : \Omega \rightarrow R$ (assuming higher values of the indicator mean better performance) is said to be *compatible* with the Pareto dominance relation if and only if:

$$I(W) > I(Z) \Rightarrow W \preceq Z \tag{20}$$

The quality indicator I is said to be *complete* if and only if:

$$W \preceq Z \Rightarrow I(W) > I(Z) \tag{21}$$

Definition 9 (*Compliance*) A quality indicator I is said to be Pareto dominance *compliant* if I is both compatible and complete with the Pareto dominance relation.

- **Error Ratio (ER)**:

This indicator is proposed by Van Veldhuizen and Lamont [37]. It corresponds to the ratio of the number of solutions that are not members of the true Pareto optimal front PF_{true} by the cardinality of the obtained solution set. Mathematically, *ER* is expressed as follows:

$$ER = \frac{\sum_{i=1}^{N} e_i}{N} \tag{22}$$

where N is the number of non-dominated solutions provided by the MOEA and $e_i = 1$ if solution i is dominated by any member from PF_{true} and $e_i = 0$ otherwise. $ER = 1$ means that no solution belongs to the true front PF_{true} and $ER = 0$ when all solutions are in the true front.

- **Set Coverage (SC)**:

This indicator can be termed relative coverage of two solution sets [34]. *SC* is defined as the mapping of the pair (*W*, *Z*) to the interval [0,1] as follows:

$$SC(W, Z) = \frac{|\{z \in Z; \exists w \in W : z \preceq w\}|}{|Z|} \quad (23)$$

SC(W,Z) expresses the percentage of solutions from *Z* that dominates solutions in *W*. *SC(W,Z)=1* means that each solution in *Z* dominates at least one solution from *W*; while *SC(W,Z)=0* means the opposite (i.e., there is no solution from *Z* dominating solutions from *W*).

- **Binary additive epsilon indicator**:

This metric takes a pair of non-dominated solution sets *W* and *Z* as inputs and returns a pair of numbers as outputs (I_W, I_Z) such that [36]:

$$I_W = I_{\varepsilon+}(W, Z) = \underset{\varepsilon \in R}{Inf}\, \{\forall z \in Z; \exists w \in W : w \preceq_{\varepsilon+} z\} \quad (24)$$

$$I_Z = I_{\varepsilon+}(Z, W) = \underset{\varepsilon \in R}{Inf}\, \{\forall w \in W; \exists z \in Z : z \preceq_{\varepsilon+} w\} \quad (25)$$

$I_{\varepsilon+}(W, Z)$ expresses the minimum quantity ε by which each solution from *W* must be translated in the objective space so that each solution from *Z* becomes dominated by (or equal to) at least one member from *W*. A pair of numbers $(I_W \leq 0, I_Z > 0)$ indicates that *W* is strictly better than *Z*, while a pair of numbers $(I_W > 0, I_Z > 0)$ means that *W* and *Z* are incomparable. Nevertheless, if I_W is less than I_Z, then in a weaker sense, we can say that *W* is better than *Z* because the minimum ε value needed so that *W* ε-dominates *Z* is smaller than the ε value needed so that *Z* ε-dominates *W*.

- **Generational distance (GD)**:

This indicator estimates how far are the elements in the Pareto front produced by the MOEA from those in the true Pareto front of the problem (i.e.,PF_{true}) [37]. It is given by the following equation:

$$GD = \frac{\sqrt{\sum_{i=1}^{N} d_i^2}}{|PF_{true}|} \quad (26)$$

where N is number of non-dominated solutions provided by the MOEA and d_i is the distance between each of these solutions to its nearest member from PF_{true}. A variant of this indicator is the *Inverted Generational Distance* (IGD) in which a reference true Pareto front is used and its elements are compared with respect to the approximation produced by the MOEA.

- **Spread (Δ)**:

The metric Δ measures the deviation among consecutive solutions in the Pareto front *PF* furnished by the MOEA [1]. Analytically, Δ is stated as follows:

$$\Delta = \sum_{i=1}^{|PF|} \frac{|dist_i - \overline{dist}|}{|PF|} \tag{27}$$

where $dist_i$ the Euclidean distance between two consecutive solutions in PF and $\overline{dist}$ is the average of these distances. In order to ensure that this calculation takes into account the spread of solutions in the entire region of the true front, the boundary solutions in the non-dominated front are included. For a perfect distribution, $\Delta = 0$ which means that $dist_i$ is constant for all i.

- **HyperVolume (HV)**:

This indicator, called also *S metric*, estimates the hypervolume of the portion of the objective space which is dominated by an approximation set [22]. The larger *HV* value is, the better the result is. This metric assesses both convergence and diversity. The *HV* indicator can be expressed as follows:

$$HV = \bigcup_i vol_i \mid i \in PF \tag{28}$$

where vol_i corresponds to hyperarea bounded by a pre-specified reference point and a solution i. The *HV* metric is *compatible* and *complete* with the Pareto dominance relation; thereby *HV* is said to be Pareto *compliant* which is an important feature for this indicator.

- **Spacing (S)**:

This metric assesses the solution distribution along the Pareto front and it is given by:

$$S = \sqrt{\frac{1}{|PF| - 1} \sum_{i=1}^{|PF|} (dis_i - \overline{dis})^2} \tag{29}$$

where $dis_i = \min_{j \in PF} \sum_{m=1}^{k} |f_m^i - f_m^j|$ and $\overline{dis}$ is the mean of these distances. The distance measure is the minimum value of the sum of the absolute differences in objective function values between solution i and any other solution in the Pareto optimal set. $S = 0$ means that all solutions are equally distributed along the Pareto front.

- **Unary R2 indicator**:

This indicator uses a set of utility functions to assess an approximation set. The unary R2 indicator can be defined as follows [38]:

$$R2(A, U) = \frac{1}{|U|} \sum_{u \in U} min_{a \in A} u(a) \tag{30}$$

where A is an approximation set and U is a set of utility functions. Different utility functions can be used such as the weighted sum and the weighted Tchebycheff function. This indicator evaluates both convergence and diversity. It is also recommended to use the R2 indicator for many-objective problems because it requires a lower computational cost in comparison to the HV indicator.

- **Chi-square-like deviation measure (χ^2-like deviation)**:

Proposed by Srinivas and Deb [19], this indicator evaluates the diversity of the obtained solution set *PF*. *PF* solutions are compared with respect to a uniformly distributed set of PF_{true} called F. For each $i \in \{1, 2, \ldots, |F|\}$, we denote by n_i the number of solutions in *PF* whose distance from i is less than a user-specified quantity ω. Then, the measure is computed as follows:

$$\chi = \sqrt{\sum_{i=1}^{|F|+1} (\frac{n_i - \overline{n_i}}{\sigma_i})^2} \tag{31}$$

The ideal distribution is achieved when all the neighborhoods of points in F have the same cardinality, i.e., if for each solution i in F there is $\overline{n_i} = \frac{|PF|}{|F|}$ points whose distance from i is less than ω, then $\chi = 0$. The variance σ_i^2 is proposed to be $\sigma_i^2 = \overline{n_i}(1 - \frac{\overline{n_i}}{|PF|})$ for all $i \in \{1, 2, \ldots, |F|\}$. The lower the χ value is, the better the distribution is.

4 Conclusion

Through this chapter, we have provided a comprehensive review of the EMO research field. We classified MOEAs based on two main criteria: (1) the use of the Pareto dominance as a selection criterion and (2) the elitism. Figure 11 illustrates a cartography of the different discussed MOEAs. Non-Elitist approaches are seen as a first generation of MOEAs while the second generation corresponds to the elitist methods. Both the use of scalarizing functions to decompose the original MOP into a collection of sub-problems and the use of a performance indicator as a selection criterion can be considered as the third generation of MOEAs. In the literature, several studies are recently conducted in those two directions. We have presented how MOEA output can be assessed by means of quality metrics and difficult test functions with predefined Pareto optimal fronts each having some geometrical features presenting challenges to every search method. As discussed through this chapter, most of the described MOEAs have shown their effectiveness and efficiency in ensuring not only convergence towards the Pareto front but also diversity between the final obtained solutions. However, this fact does not resolve the problem of decision making since

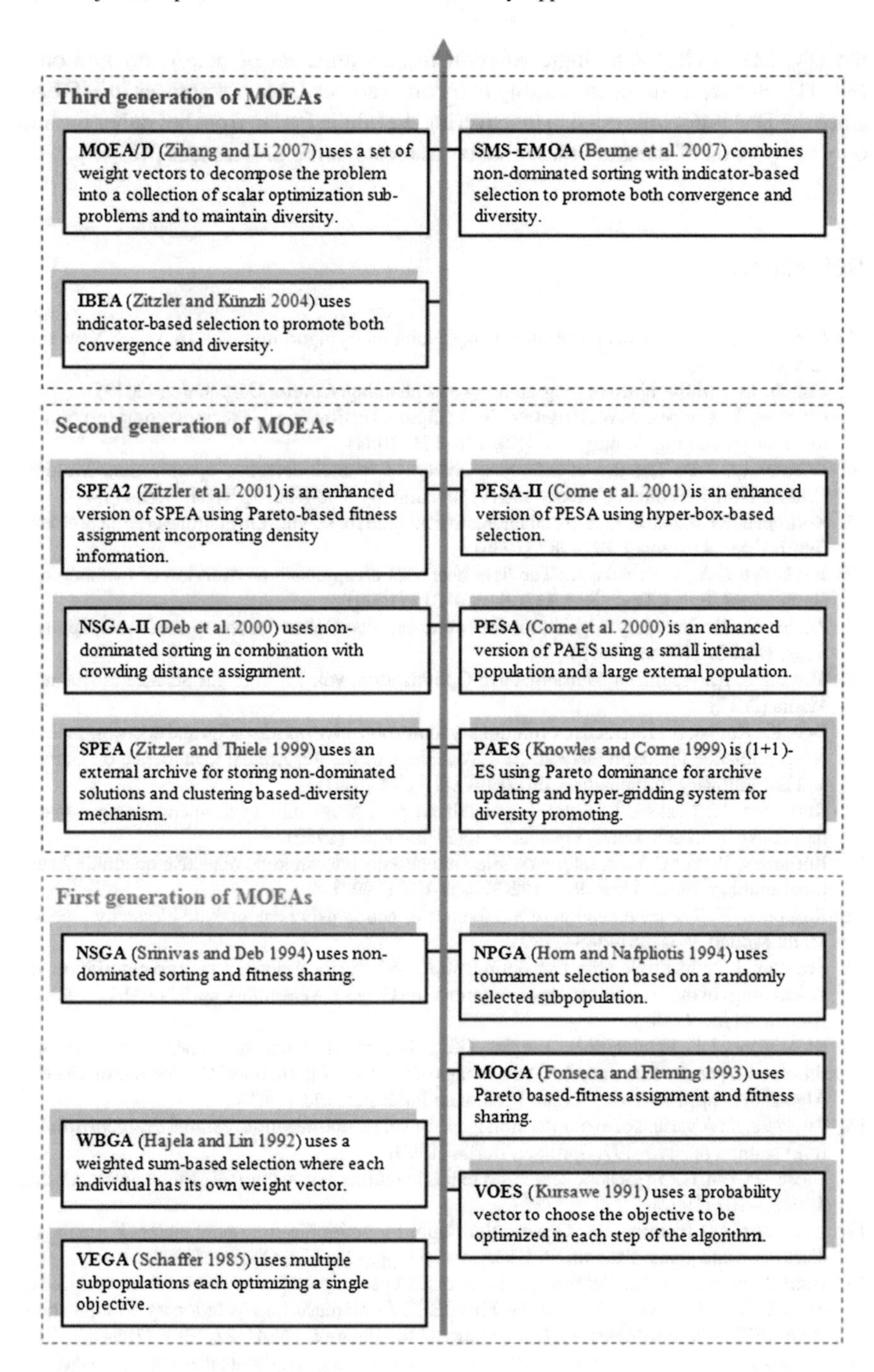

Fig. 11 MOEA cartopgraphy

the DM has to choose a single solution from a huge set of non-dominated ones [39–41]. Hence, it will be interesting to incorporate the DMs preferences in MOEAs since the DM is not interested in discovering the whole Pareto front but rather finding only the portion of the front that matches at most his/her preferences [42, 43].

References

1. Deb, K.: Multi-objective optimization using evolutionary algorithms, vol. 16. Wiley, New York (2001)
2. Cohon, J.L.: Multiobjective programming and planning. Courier Corporation (2013)
3. Charnes, A., Cooper, W.W., Ferguson, R.O.: Optimal estimation of executive compensation by linear programming. Manag. Sci. **1**(2), 138–151 (1955)
4. Wierzbicki, A.P.: The use of reference objectives in multiobjective optimization. Multiple Criteria Decision Making Theory and Application, pp. 468–486. Springer, Berlin (1980)
5. Korhonen, P.J., Laakso, J.: A visual interactive method for solving the multiple criteria problem. Eur. J. Oper. Res. **24**(2), 277–287 (1986)
6. Jaszkiewicz, A., Słowiński, R.: The light beam search approach-an overview of methodology applications. Eur. J. Oper. Res. **113**(2), 300–314 (1999)
7. Zitzler, E.: Evolutionary Algorithms for Multiobjective Optimization: Methods and Applications, vol. 63. Citeseer (1999)
8. Miettinen, K.: Nonlinear Multiobjective Optimization, vol. 12. Springer Science & Business Media (2012)
9. Deb, K., Kumar, A.: Interactive evolutionary multi-objective optimization and decision-making using reference direction method. In: Proceedings of the 9th Annual Conference on Genetic and Evolutionary Computation, pp. 781–788. ACM (2007)
10. Korhonen, P., Laakso, J.: Solving generalized goal programming problems using a visual interactive approach. Eur. J. Oper. Res. **26**(3), 355–363 (1986)
11. Korhonen, P., Yu, G.Y.: A reference direction approach to multiple objective quadratic-linear programming. Eur. J. Oper. Res. **102**(3), 601–610 (1997)
12. Korhonen, P.: The specification of a reference direction using the analytic hierarchy process. Math. Modell. **9**(3), 361–368 (1987)
13. Schaffer, J.D.: Multiple objective optimization with vector evaluated genetic algorithms. In: Proceedings of the 1st International Conference on Genetic Algorithms, pp. 93–100. L. Erlbaum Associates Inc. (1985)
14. Richardson, J.T., Palmer, M.R., Liepins, G.E., Hilliard, M.: Some guidelines for genetic algorithms with penalty functions. In: Proceedings of the Third International Conference on Genetic Algorithms, pp. 191–197. Morgan Kaufmann Publishers Inc. (1989)
15. Kursawe, F.: A variant of evolution strategies for vector optimization. Parallel Problem Solving from Nature, pp. 193–197. Springer, Berlin (1990)
16. Hajela, P., Lin, C.-Y.: Genetic search strategies in multicriterion optimal design. Struct. Optim. **4**(2), 99–107 (1992)
17. Fonseca, C.M., Fleming, P.J.: Genetic algorithms for multiobjective optimization: Formulation discussion and generalization. In: ICGA, vol. 93, pp. 416–423, Citeseer (1993)
18. Horn, J., Nafpliotis, N., Goldberg, D.E.: A niched pareto genetic algorithm for multiobjective optimization. In: Proceedings of the First IEEE Conference on Evolutionary Computation, 1994. IEEE World Congress on Computational Intelligence, pp. 82–87, IEEE (1994)
19. Srinivas, N., Deb, K.: Muiltiobjective optimization using nondominated sorting in genetic algorithms. Evol. Comput. **2**(3), 221–248 (1994)
20. Holland, J.H.: Adaptation in Natural and Artificial Systems: An Introductory Analysis with Applications to Biology, Control, and Artificial Intelligence. U Michigan Press (1975)

21. Bechikh, S., Chaabani, A., Said, L.B.: An efficient chemical reaction optimization algorithm for multiobjective optimization. IEEE Trans. Cybern. **45**(10), 2051–2064 (2015)
22. Zitzler, E., Thiele, L.: Multiobjective evolutionary algorithms: a comparative case study and the strength pareto approach. Evol. Comput. Trans. IEEE **3**(4), 257–271 (1999)
23. Zitzler, E., Laumanns, M., Thiele, L.: Spea2: Improving the Strength Pareto Evolutionary Algorithm (2001)
24. Deb, K., Agrawal, S., Pratap, A., Meyarivan, T.: A fast elitist non-dominated sorting genetic algorithm for multi-objective optimization: Nsga-ii. In: Parallel Problem Solving from Nature PPSN VI, pp. 849–858. Springer, New York (2000)
25. Deb, K., Pratap, A., Agarwal, S., Meyarivan, T.: A fast and elitist multiobjective genetic algorithm: Nsga-ii. IEEE Trans. Evol. Comput. **6**(2), 182–197 (2002)
26. Knowles, J., Corne, D.: The pareto archived evolution strategy: A new baseline algorithm for pareto multiobjective optimisation. In: Proceedings of the 1999 Congress on Evolutionary Computation, 1999. CEC 99, vol. 1. IEEE (1999)
27. Knowles, J.D., Corne, D.W.: Approximating the nondominated front using the pareto archived evolution strategy. Evol. Comput. **8**(2), 149–172 (2000)
28. Corne, D.W., Knowles, J.D., Oates, M.J.: The pareto envelope-based selection algorithm for multiobjective optimization. In: Parallel Problem Solving from Nature PPSN VI, pp. 839–848. Springer (2000)
29. Corne, D.W., Jerram, N.R., Knowles, J.D., Oates, M.J., et al.: Pesa-ii: Region-based selection in evolutionary multiobjective optimization. In: Proceedings of the Genetic and Evolutionary Computation Conference (GECCO 2001). Citeseer (2001)
30. Zitzler, E., Künzli, S.: Indicator-based selection in multiobjective search. In: Parallel Problem Solving from Nature-PPSN VIII, pp. 832–842. Springer (2004)
31. Beume, N., Naujoks, B., Emmerich, M.: Sms-emoa: Multiobjective selection based on dominated hypervolume. Eur. J. Oper. Res. **181**(3), 1653–1669 (2007)
32. Azzouz, N., Bechikh, S., Said, L.B.: Steady state ibea assisted by mlp neural networks for expensive multi-objective optimization problems. In: Proceedings of the 2014 Conference on Genetic and Evolutionary Computation, pp. 581–588. ACM (2014)
33. Zhang, Q., Li, H.: Moea/d: A multiobjective evolutionary algorithm based on decomposition. IEEE Trans. Evol. Comput. **11**(6), 712–731 (2007)
34. Zitzler, E., Deb, K., Thiele, L.: Comparison of multiobjective evolutionary algorithms: Empirical results. Evol. Comput. **8**(2), 173–195 (2000)
35. Deb, K., Thiele, L., Laumanns, M., Zitzler, E.: Scalable multi-objective optimization test problems. In: Proceedings of the 2002 Congress on Evolutionary Computation, 2002. CEC'02, vol. 1, pp. 825–830. IEEE (2002)
36. Zitzler, E., Thiele, L., Laumanns, M., Fonseca, C.M., Da Fonseca, V.G.: Performance assessment of multiobjective optimizers: an analysis and review. IEEE Trans. Evol. Comput. **7**(2), 117–132 (2003)
37. Van Veldhuizen, D.A., Lamont, G.B.: On measuring multiobjective evolutionary algorithm performance. In: Proceedings of the 2000 Congress on Evolutionary Computation, 2000, vol. 1, pp. 204–211. IEEE (2000)
38. Hernández Gómez, R., Coello Coello, C.A.: Improved metaheuristic based on the r2 indicator for many-objective optimization. In: Proceedings of the 2015 on Genetic and Evolutionary Computation Conference, pp. 679–686. ACM (2015)
39. Bechikh, S., Said, L.B., Ghédira, K.: Negotiating decision makers' reference points for group preference-based evolutionary multi-objective optimization. In: 2011 11th International Conference on Hybrid Intelligent Systems (HIS), pp. 377–382. IEEE (2011)
40. Bechikh, S., Said, L.B., Ghédira, K.: Group preference based evolutionary multi-objective optimization with nonequally important decision makers: Application to the portfolio selection problem. Int. J. Comput. Inf. Syst. Ind. Manag. Appl. **5**(278–288), 71 (2013)
41. Kalboussi, S., Bechikh, S., Kessentini, M., Said, L.B.: Preference-based many-objective evolutionary testing generates harder test cases for autonomous agents. In: Search Based Software Engineering, pp. 245–250. Springer (2013)

42. Bechikh, S., Kessentini, M., Said, L.B., Ghédira, K.: Chapter four-preference incorporation in evolutionary multiobjective optimization: A survey of the state-of-the-art. Adv. Comput. **98**, 141–207 (2015)
43. Bechikh, S.: Incorporating decision makers preference information in evolutionary multi-objective optimization. PhD thesis, University of Tunis, ISG-Tunis, Tunisia (2013)

Dynamic Multi-objective Optimization Using Evolutionary Algorithms: A Survey

Radhia Azzouz, Slim Bechikh and Lamjed Ben Said

Abstract Dynamic Multi-objective Optimization is a challenging research topic since the objective functions, constraints, and problem parameters may change over time. Although dynamic optimization and multi-objective optimization have separately obtained a great interest among many researchers, there are only few studies that have been developed to solve Dynamic Multi-objective Optimisation Problems (DMOPs). Moreover, applying Evolutionary Algorithms (EAs) to solve this category of problems is not yet highly explored although this kind of problems is of significant importance in practice. This paper is devoted to briefly survey EAs that were proposed in the literature to handle DMOPs. In addition, an overview of the most commonly used test functions, performance measures and statistical tests is presented. Actual challenges and future research directions are also discussed.

Keywords Dynamic optimization · Multi-objective optimization · Evolutionary algorithms · Survey · Real world applications · Test functions · Performance metrics · Statistical Tests

1 Introduction

In addition to the need for simultaneously optimizing several competing objectives, many real-world problems are also dynamic in nature. These problems are called DMOPs and they are characterized by time-varying objective functions and/or constraints. Thus, the optimization goal is not only to evolve a near-optimal PF, but also

R. Azzouz (✉) · S. Bechikh · L. Ben Said
SOIE Lab, Computer Science Department, ISG-Tunis, University of Tunis,
Bouchoucha City, 2000 Le Bardo, Tunis, Tunisia
e-mail: azouzradhia@gmail.com

S. Bechikh
e-mail: slim.bechikh@gmail.com

L. Ben Said
e-mail: lamjed.bensaid@isg.rnu.tn

S. Bechikh et al. (eds.), *Recent Advances in Evolutionary Multi-objective Optimization*, Adaptation, Learning, and Optimization 20,
DOI 10.1007/978-3-319-42978-6_2

to continually and rapidly discover the desired one before the next change occurs. Applying EAs to solve dynamic optimization problems has obtained great attention among many researchers. However, most of existing works are restricted to the single-objective case. To the best of our knowledge, the earliest application of EAs to dynamic environments dates back to 1966 [1]. However, it was not until the late 1980s that the subject becomes a research topic. Although many other optimization techniques have been adapted to dynamic environments such as the particle swarm optimization [2] and the artificial immune systems [3, 4], the EA area is still the largest one. When dealing with DMOPs, the EA should be able not only to evolve a near-optimal and diverse PF, but also to continually track time-changing environment. In fact, two ways exist to react to a change of the environment: (1) to consider each change as the arrival of a new optimization problem that has to be solved from scratch or (2) to use knowledge about the previous search in order to accelerate optimization after a change. The first approach is not always applicable due to a time limit [5]. In the second case, the optimization algorithm has to ensure adaptability since convergence during the run may cause a lack of diversity. Such goal of adaptability and track of the optimal PF implies a conflicting requirement of convergence and diversity. There are few works handling DMOPs which include diversity introduction-based approaches [6, 7], change prediction-based approaches [8, 9], memory-based approaches [10], and parallel approaches [11].

The topic of dynamic optimization was reviewed in the past but this has mainly covered dynamic single-objective optimization [5, 12, 13]. The research field of dynamic multi-objective optimization is an emerging area in evolutionary computation ant it attracts more and more researchers. This is why, it becomes primordial to have a look on what has been done in the past and what could be done in the future. Only a few number of works reviewing dynamic multi-objective optimization topic exist in the literature like [2, 14, 15]. This paper is proposed as a step towards fulfilling this gap. It is mainly devoted to briefly survey EAs proposed for handling DMOPs and to present a repository about the most commonly used dynamic multi-objective benchmark functions and performance measures.

Section 2 highlights the most important definitions related to this area. In Sect. 3, a number of classifications of dynamic environments are presented. Section 4 provides an overview of the most important works that deal with the problematic of the use of EAs to handle DMOPs. Advantages and shortcomings of different approaches are outlined. Section 5 presents the most commonly used test problems on assessing the performance of dynamic EAs while Sect. 6 explains the performance metrics and statistical tests used when comparing different dynamic approaches. A discussion part is presented in Sect. 7. Finally, Sect. 8 concludes this paper and gives some suggestions for future research.

2 Definitions

Definition 1 Dynamic Multi-objective Optimization Problem.

A DMOP can be defined as the problem of finding a vector of decision variables $x(t)$, that satisfies a restriction set and optimizes a function vector whose scalar values represent objectives that change over time. Considering a minimization problem, the DMOP can be formally defined as follows:

$$\begin{aligned} Min\ F(x,t) &= \{f_1(x,\, t),\ f_2(x,\, t), \ldots,\ f_M(x,\, t)\}\backslash x \in X^n \\ s.t.\ g(x,\, t) &> 0,\ h(x,\, t) = 0 \end{aligned} \tag{1}$$

where x is the vector of decision variables; f is the set of objectives to be minimized with respect to time. The functions of g and h represent respectively the set of inequality and equality constraints while t represents the time or the dynamic nature of the problem and M represents the number of objectives to be minimized.

Definition 2 Dynamic Pareto Optimal solution.

A decision vector $x^*(i,\, t)$ is said to be a Pareto optimal solution if there is not any other feasible decision vector, $x\,(j,\, t)$ such that

$$f(j,\, t) \prec f(i,\, t)^* \backslash\ f(j,\, t) \in F^M$$

Where $\prec$ represents the Pareto dominance relation.

Definition 3 Dynamic Optimal Pareto Front.

The optimal PF at time t, denoted as $PF(t)^*$, is the set of Pareto optimal solutions with respect to the objective space at time t such that

$$PF(t)^* = \{f(i,\, t)^* |\ \nexists\, f(j,\, t) \prec f(i,\, t)^*,\ \ f(j,\, t) \in F^M\}$$

Definition 4 Dynamic Pareto Optimal Set.

The Pareto-optimal set at time t, denoted as $PS(t)^*$, is the set of Pareto optimal solutions with respect to the decision space such that

$$PS(t)^* = \{x_i^* |\ \nexists\, f(x_j,\, t) \prec f(x_i^*,\, t)^*,\ \ f(x_j,\, t) \in F^M\}$$

Definition 5 Change Severity.

The change severity signifies how fundamental the changes are in terms of their magnitude. It measures the relative strength of the landscape change by comparing the landscape before and after a change [16].

Definition 6 Change Frequency.

The change frequency determines how often the environment changes. Usually it is measured as the number of generations or the number of fitness functions evaluations from one landscape change to the next [16].

3 Classifications of DMOPs

A number of classifications have been proposed in the literature based on the frequency, severity, and predictability of changes.

- Frequency-based classification: When the change frequency increases, the time dedicated for adaptation becomes shorter which makes the problem more difficult.
- Severity-based classification: The change severity (rate) defines its degree. There can be a large change in the problem or there can be a small change. It is easier for the algorithm in the second case to converge to the optimal PF since information gained from the previous environment can be exploited and reused to accelerate the convergence speed. If the change severity is large, each instance of the problem may be completely unrelated to the next one. Thus, it may be useful to completely re-start the algorithm.
- Predictability-based classification: The change predictability indicates its regularity. A change is random when it is independent of the previous one while it is considered non-random or predictable when it is deterministic. This class could be divided into cyclic changes (changes are periodic) or acyclic ones.
- Classification based on the relation between the optimal PF and the optimal PS: Farina et al. [17] identified four different types of DMOPs according to changes affecting the optimal PF and the optimal PS as follows:
  - type I, where the optimal PS (PS^*) changes while the optimal PF (PF^*) remains invariant;
  - type II, where both PS^* and PF^* change;
  - type III, where PF^* changes while PS^* remains invariant; and
  - type IV, where both PS^* and PF^* remain invariant.

    Farina et al. noted that, even if PS^* and PF^* remain unchanged in Type IV problems, other regions of the fitness landscape can be changing. It is the case when for example only the local optima vary over time. These four types are summarised in Table 1.

Table 1 Dynamic multi-objective optimization environment types

$PF(t)^*$	$PS(t)^*$	
	No change	Change
No change	Type VI	Type I
Change	Type III	Type II

4 Dynamic Multi-objective Optimization Using EAs

Dynamic optimization problems include Dynamic Single-objective Optimization Problems (DSOPs) and DMOPs. EAs were first applied to DSOPs. In fact, the optimization algorithm has to ensure adaptability since convergence during the run may cause a lack of diversity. Thus, the algorithm loses its ability to flexibly react to changes. For this reason, several additional mechanisms were proposed to keep diversity in the population. Diversity can be either maintained throughout the run [18, 19], or increased after a change detection by taking explicit actions such as reinitialization or hypermutation [20, 21]. Also, many other approaches have been proposed such as memory-based approaches [22, 23], multipopulation approaches [24, 25], predictive approaches [26, 27], etc. A number of interesting surveys of these approaches exist in the literature. Interested readers may refer to [5].

The main difficulty in the multi-objective case is that the PF of a DMOP may change when the environment changes which makes the task of optimization more difficult. Contrarily to the single-objective case, there are few works dealing with DMOPs. As well, the number of papers presenting an overview of existing approaches is very limited. This is why, we devote this chapter to briefly survey EAs for handling DMOPs for which we propose the following classification.

4.1 Diversity-Based Approaches

4.1.1 The Dynamic Non-dominated Sorting Genetic Algorithm II (D-NSGA-II)

A conflicting requirement of convergence and diversity is imposed when dealing with DMOPs since convergence during the run may cause a lack of diversity which may cause that the algorithm loses its ability to adapt and flexibly react to changes. One way to deal with this issue is to increase diversity after detecting a change. Another way is to try to maintain a good level of diversity all over the search process. One important work belonging to this category of approaches is Dynamic NSGA-II (DNSGA-II) proposed in 2006 [6] where Deb et al. extended NSGA-II to handle DMOPs by introducing diversity at each change detection. In fact in each generation, few solutions are randomly selected and re-evaluated. If there is a change in the objectives or constraint violation values, the problem is considered to be changed. Then, all outdated solutions (i.e., parent solutions) are re-evaluated. This process allows both offspring and parent solutions to be evaluated using the changed objectives and constraints functions. Two versions of the proposed dynamic NSGA-II were suggested. Diversity is introduced in the first version (DNSGA-II-A) through the replacement of $\zeta\%$ of the new population with new randomly created solutions. In the second version (DNSGA-II-B), diversity is ensured by replacing $\zeta\%$ of the new population with mutated solutions. Authors also suggest a decision-making aid to help identify one dynamic single optimal solution on-line. One of the merits of this work is that it can also solve constrained DMOPs. This work has been evaluated on a modified version of the FDA2 test problem and a real world optimization of a hydro-thermal power scheduling problem involving two conflicting objectives. The dynamicity of this problem is due to the change in demand in power with time [6]. The first version based on random initialization has demonstrated better performance on problems subjected to a large change while, the second version performs well on problems undergoing a small change in the problem. Nevertheless, both versions are sensitive to the choice of the population ratio ζ and the change frequency.

4.1.2 The Dynamic Constrained NSGA-II (DC-NSGA-II)

In [7], authors proposed an adaptation of DNSGA-II 1 to deal with dynamic constraints by replacing the used constraint-handling mechanism by a more elaborated and self-adaptive penalty function. The resulting algorithm is called Dynamic Constrained NSGA-II (DC-NSGA-II). Moreover, to fill the gap of the lack of benchmarks that simultaneously take into account the dynamicity of objective functions and constraints, authors also proposed a set of test problems that extend the CTPs suite of static constrained multi-objective problems [28]. The new dynamic constrained MOPs denoted as Dynamic CTPs (i.e., DCTPs) present different challenges to the optimization algorithm since the PF, the PS and the constraints change simultaneously over time. In fact, DNSGA-II uses the constraint dominance principle used in NSGA-II to deal with constraints. However, since this principle prefers feasible solutions over infeasible ones, it often results in a premature convergence due to the loss of diversity over time. This is why, authors proposed to replace the dominance principle used to handle constraints by the penalty function proposed in [29]. They supposed that the constraint-handling technique should be able to find feasible individuals and to maintain some infeasible solutions allowing to avoid premature convergence; while the dynamic EA would be able to ensure the diversity in the population and to track changing PFs. Furthermore, the diversity introduction mechanism was ameliorated. A feasibility condition was added before incorporating any random or mutated solution into the population, since accepting infeasible solutions may slow down convergence. This work has been evaluated on the proposed DCTPs problems where it was able to handle dynamic environments and to track changing PFs with time-varying constraints. Moreover, the obtained results have demonstrated the advantages of this algorithm over the original DNSGA-II versions on both aspects of convergence and diversity. However, this approach faces difficulties when dealing with problems having many local optimal PFs.

4.1.3 Individual Diversity Multi-objective Optimization EA (IDMOEA)

Chen et al. [30] proposed to explicitly maintain genetic diversity by considering it as an additional objective in the optimization process. They presented the individual

Table 2 Diversity-based dynamic EAs

Algorithm	Compared to	Used benchmarks	Used performance metrics
D-NSGA-II [6]	–	A modified version of FDA2 [6] and the hydro-thermal power scheduling problem [6]	HyperVolume (HV) ratio [31]
DC-NSGA-II [7]	D-NSGA-II [6]	DCTPs test problems [7]	Inverted Generational Distance (IGD) [32], HV ratio [31], and MS [10]
IDMOEA [30]	–	FDA1 and FDA5 [17]	GD [33] and entropy [30]

diversity multi-objective optimization EA (IDMOEA) that uses a new diversity preserving evaluation method which is called Individual Diversity Evolutionary Method (IDEM). The goal of IDEM is to add a useful selection pressure addressed towards both the optimal PS and the maintenance of diversity [30]. The average of individual's entropy is used as a diversity measure. The first step of IDMOEA is to verify if there is a change in the environment. If an environmental change takes place, a new population is created using the best individuals of the current population and the archive. Otherwise, the new population is created as a copy of the current population. Then, binary tournament selection is executed to select parents on which the crossover will be performed. Mutation is applied on the produced offsprings and the population and archive update are performed to maintain elite solutions. The archive is updated by adding non-dominated individuals of the population to it. If the archive attends its maxsize, individuals with better diversity are maintained. The performance of IDMOEA was evaluated on FDA1 and FDA5 [17]. The results showed that the algorithm is effective at converging towards the optimal PS and to track changing PFs while maintaining a diverse set of solutions.

Table 2 summarizes the algorithms discussed in this section, and the algorithms that they were confronted to, as well as the benchmark functions and the performance measures that they were evaluated on.

4.2 *Change Prediction-Based Approaches*

To exploit past information and anticipate the location of the new optimal solutions, a prediction model may be used when the behavior of the dynamic problem follows a certain trend. In fact, these approaches are used to reduce the number of functions evaluations while reserving the quality of optimized solutions. This is by predicting the location of the new optimal PF or the new optimal PS based on informations about previous environments.

4.2.1 Dynamic Queuing Multi-objective Optimizer (D-QMOO)

Hatzakis and Wallace [8] proposed a forecasting technique called Feed-forward Prediction Strategy (FPS) in order to estimate the location of the optimal PS. Then an anticipatory population called a prediction set is placed in the neighborhood of the forecast in order to accelerate the discovery of the next PS. Since this work deals with only bi-objective optimization problems, this set is formed by selecting two anchor points (i.e., the extreme solutions in the obtained PF: min (f_1) and min (f_2)) as vertices and tracking and predicting them as next-step optima. In fact, for each point, the sequence of the past optimal locations is used as input to a forecasting model, which produces an estimate for the next location. As soon as the next time step arrives and the objective functions change, the prediction set is inserted into the population. If the prediction is successful, the predicted individuals will accelerate the convergence of the rest of the population and help the discovery of the next PS. Since the prediction might be unsuccessful, or the temporal change pattern might not be identifiable by the forecasting method, the use of the prediction strategy can not be sufficient to tackle with the dynamicity of the problem. Authors suggested the use of a convergence/diversity balance technique. It consists in composing the total population at the beginning of the optimization of three parts: (1) the prediction set, (2) the non-dominated front and (3) the cruft (i.e., the dominated set) whose function is to preserve diversity and to handle any unpredictable change. The FPS was combined with the EA developed by Leyland based on the Queuing Multi-objective Optimizer (QMOO) [8]. The resulting algorithm is called Dynamic QMOO (D-QMOO). The main advantage of this algorithm is that instead of re-introducing past optimal solutions into the evolving population, information is exploited to predict future behavior of the dynamic problem, aiming at a faster convergence to the new PF. Nevertheless, only one dynamic test problem, which is the FDA1 problem [17], is considered to examine its performance, while the precision of the prediction should be further improved.

4.2.2 The Work of Hatzakis and Wallace (2006)

This work presents an extension to the FPS proposed in [8] where authors have studied the influence of the size and the distribution of the anticipatory population on the search performance. Since the prediction almost have an amount of error due to the accuracy of the optimal solution's history and the accuracy of the forecasting model, it is important to populate the forecast neighborhood instead of only placing a single individual on the forecasted coordinates. To deal with this issue, authors have proposed to create prediction sets in the form of a hypercube around the forecast coordinates, dimensioned in proportion to the expected forecast error. An individual is placed at the center and at each hypercube corner. The main disadvantage of a hypercube topology is its computational cost whenever the dimension of the design vector increases. Thus, authors have proposed to use a two-level Latin hypercube with 3 individuals per prediction point: the centre point and the two LH points. On the other hand, since in [8], only the two anchor points were selected to be tracked and forecasted, this approach leaves a large portion of the PS uncovered mostly when its shape is non-linear and complex. In this work, an intermediate point defined as the non-dominated solution closest to the ideal point and called CTI (Closest-To-Ideal) is proposed to be selected together with the extremities of the PF. The proposed topologies of populating the neighborhood of the forecast were tested on the FDA1 test problem [17]. Results have shown that the hypercube has the best accuracy (least error) for low dimension design vectors, while, the Latin hypercube has the best accuracy with 6 decision variables. Moreover, initial experiments show that including the CTI point in the prediction set improves performance mostly with a high change frequency. Nevertheless, selecting the CTI point may be difficult when the front is very concave and large parts of it have almost the same distance to the ideal point.

4.2.3 The Dynamic Multi-objective EA with Predicted Re-Initialization (DMEA/PRI)

Unlike the extended FPS where only three points of the PS (the two anchor points and the CTI point) are tracked and predicted, Zhou et al. [34] proposed to predict the new locations of a number of Pareto solutions in the decision space once a change is detected. Then, individuals in the re-initialized population are generated around these predicted points. Two strategies for population re-initialization were introduced. The first strategy is to predict the new individuals' locations from the previous locations changes. The population is then partially or completely replaced by the new individuals generated based on prediction. The second strategy is to add to the population a "predicted" Gaussian noise whose variance is estimated according to previous changes. A framework of the dynamic multi-objective EA with predicted re-initialization called (DMEA/PRI) and based on predicted re-initialization strategies was presented. Moreover, four methods for re-initialization have been studied and compared: (1) random re-initialization method (RND) such that initial populations are randomly generated, (2) variation method (VAR) using the variation with a predicted noise strategy, (3) prediction method (PRE) where the new individuals are generated around the predicted locations and (4) a hybrid method (V&P) where half of population is generated by method 2 and half is created by method 3. The performance of the proposed methods was assessed on two test problems: FDA1 [17] and ZJZ which is a modified version of FDA1 using the method proposed in [35] in order to take into account a linear linkage between decision variables. The empirical results have shown that for the FDA1 test problem, the RND method does not work at all. The VAR method does not perform well while the V&P method and the PRE method are comparable and perform better than the RND and VAR strategies. For the ZJZ problem, when the time window increases, the V&P and PRE methods outperform other ones.

4.2.4 The Work of Roy and Mehnen (2008)

In [36], Roy and Mehnen introduced a dynamic multi-objective EA using forecasting and desirability functions. In fact, the proposed algorithm is an adaptation of DNSGA-II [6] where diversity is no more introduced when a change occurs by adding some random or mutated solutions. Instead, parent population is discarded and only offspring individuals are re-evaluated before that the algorithm restart. The objective functions are transformed using desirability functions to guide the search towards the most interesting parts of the optimal PF according to an expert or decision maker's preferences. Moreover, a forecasting is incorporated into the algorithm. It consists on segmenting the objective space into a grid of hyper-cubes. Each cube of the grid represents a section of the PF for a certain time t. At each time t, representative points of each cube are determined and a two dimensional time series is assigned to each one. Then for each objective, a state space model is used for modelling the multi-variate timeseries. The proposed dynamic NSGA-II uses after a predefined number of generations a k forecasted values for k iterations. During these k iterations no function evaluations are performed. The algorithm was tested on a real-world problem of machining of material with continuously varying properties, also known as the gradient material problem. The results indicated that the use of desirability functions strongly reduce the number of obtained non-dominated solutions. Moreover, authors claimed that the multivariate analysis of more than four time series at a time resulted in forecasts with poor confidence intervals.

4.2.5 The Dynamic Multi-objective Evolutionary Gradient Search (Dynamic MO-EGS)

A new prediction strategy called dynamic predictive gradient strategy is proposed in [9] to predict the good search direction and the magnitude of changes in the decision space. Besides, a new memory technique requiring few evaluations is introduced to exploit any periodicity in the dynamic problem. Then, both techniques are incorporated into a dynamic variant of the Multi-objective Evolutionary Gradient Search (MO-EGS). The dynamic predictive gradient strategy consists in defining a set of vectors called predictive gradients relating the obtained solutions for the previous landscapes and describing the direction and the magnitude of the next change in the location of the optimal PS. The predictive gradient is used to update some individuals of the population which will guide the rest of the population towards the new optimal PS. MO-EGS is a memetic MOEA that extends the concept of Evolutionary Gradient Search for MO optimization. In order to preserve elitism, MO-EGS maintains an external archive to store the non-dominated solutions found. The gradient information of each solution needed for the estimation of the global gradient is represented by the fitness of the solution which is calculated using an aggregation function that combines the objective values of the solution into a scalar value. An implementation to adapt MO-EGS for dynamic MO optimization, called dMO-EGS, was proposed based on the dynamic predictive gradient strategy and a new selective memory technique. This technique is based on storing the outdated archive by storing only its geometric centroid and centroid variance. Moreover, to detect environment changes, few solutions are randomly selected and re-evaluated. If there is a change in the objective values, the problem is considered to be changed. To assess the performance of this algorithm, two sets of experiments were conducted on static and dynamic environments. When resolving static test problems, the proposed approach was compared to NSGA-II, SPEA2 and PAES. The results have shown that all algorithms have similar performance. On the other hand, the performance of dMO-EGS was compared to two dynamic MOEAs (i.e., dCCEA and dPAES: the dynamic version of PAES) where the same dynamic handling techniques used in dMO-EGS were implemented in dCCEA and dPAES. The results indicated that the prediction strategy is able to improve performance on all used test problems.

4.2.6 The Dynamic Multi-objective EA with Core Estimation of Distribution

In [37], Liu has proposed a Dynamic Multi-objective EA with Core Estimation of Distribution (CDDMEA) that incorporates a core estimation of distribution model to predict the location of Pareto optimal solutions of the next environment. In fact, the core of the different optimal PSs at different time steps is calculated as the average solution of each one using the mean value of each variable space dimension. Then, when a change occurs, the re-initialized population is obtained by adding the difference between the core solutions at time $t - 1$ and time $t - 2$ to each solution at time t to obtain the new solution at time $t + 1$. The performance of CDDMEA was evaluated on a test problem defined in [38] and the FDA2 test function [17] and it was compared to DNSGA-II-A [6]. Visual comparisons of the plots of the obtained PFs were performed in addition to the evaluation of the U-measure to evaluate the diversity of the obtained solutions. The authors claimed that CDDMEA is better than DNSGA-II-A but more experiments on different benchmark functions and using different performance measures still are needed. Moreover, as noted in [39], this prediction approach is based on the Pareto optimal solutions which may induce that errors in previously found optimal PS may cause the algorithm to lose track of the changing optimal solutions.

4.2.7 The Population Prediction Strategy (PPS)

More recently in 2014, Zhou et al. [40] proposed to predict a whole population rather than predicting some isolated points for continuous DMOPs. This approach, called Population Prediction Strategy (PPS) consists in dividing the PS into two parts: a center point and a manifold. When a change is detected, the next center point is predicted using a sequence of center points maintained all over the search progress, and the previous manifolds are used to estimate the next manifold. Then, PPS initializes the whole population by combining the predicted center and the estimated manifold. The center points $x^0, x^1, \ldots, x^t$ form a time series. A univariate autoregression (AR) model was applied to forecast the location of the next center x^{t+1}. For the approximation of the PS manifold C at time $t+1$, PPS records the last two approximated manifolds C^t and C^{t-1}. In fact, each point $x^t \in C^t$ is used to estimate a new point x^{t+1}. The performance of PPS was evaluated by confronting three instances of RM-MEDA [27]: (1) RM-MEDA including PPS, (2) RM-MEDA including a random initialization strategy and (3) RM-MEDA including FPS. These comparisons were done on a variety of DMOPs: FDA1 [17], FDA4 [17], dMOP1 [10], dMOP2 [10] and 4 newly proposed test functions [40]. Statistical results have demonstrated the effectiveness of this approach. Moreover, authors studied the influences of some problem parameters, the influences of different MOEA optimizers and the influences of several time series predictors. Results have shown that PPS is more suitable to linear models than nonlinear ones. Compared to the FPS, PPS has the advantages to predict a whole population instead of some isolated points with a better time and space complexities.

4.2.8 Dynamic Multiobjective EA with ADLM Model (DMOEA/ADLM)

In 2014, a new prediction model [41] has been defined to solve DMOPs with Translational optimal PS (DMOP-TPS). DMOP-TPS is a specific kind of DMOPs where the PS translates regularly over time.

Definition 7 DMOP-TPS
Let $PS(t)$ and $PS(t+1)$ be two consecutive optimal PSs at time t and $t+1$ respectively, $A(t) = (a_1(t), a_2(t), \ldots, a_n(t))$ a n-dimensional vector, a DMOP is called a DMOP-TPS if and only if for any decision variable $X^t = (x_1^t, x_2^t, \ldots, x_n^t) \in PS(t)$, there must be a decision variable $X^{t+1} = (x_1^{t+1}, x_2^{t+1}, \ldots, x_n^{t+1}) \in PS(t+1)$ which satisfies the constraints $\{x_1^{t+1} = x_1^t + a_1(t), x_2^{t+1} = x_2^t + a_2(t), \ldots, x_n^{t+1} = x_n^t + a_n(t)\}$.

When an environmental change is detected using the strategy proposed by Deb et al. [6], the population is re-initialized according to the nature of the DMOP. In fact, some new predicted individuals are generated and inserted into the current population. Taking into account the mathematical properties of a DMOP-TPS, ADLM which is a linear model inspired by the prediction strategies described in [8, 34] is designed and adopted to predict the location of these solutions. ADLM is then integrated into a basic Dynamic Multi-objective EA (DMOEA). The resulting algorithm, called DMOEA/ADLM was compared against three traditional prediction models which are MM, VARM and PREM. Experiments were conducted on six DMOP-TPS test problems (FDA1 and FDA5 and their extensions FDA1E, FDA1L, FDA5E and FDA5L) [41]. Simulation results have shown the superiority of the proposed model over the rest of the prediction models on both aspects of convergence and time complexity.

4.2.9 The Kalman Filter Assisted MOEA/D-DE Algorithm (MOEA/D-KF)

Muruganantham et al. [42] proposed a dynamic multi-objective EA that uses a Kalman Filter-based prediction model. Whenever a change is detected, Kalman Filter is applied to the whole population to direct the search towards the new Pareto optimal solutions in the decision space. The proposed algorithm is based on the Multiobjective EA with Decomposition based on Differential Evolution (MOEA/D-DE) and is called Kalman Filter prediction based DMOEA (MOEA/D-KF). This work was tested on the IEEE CEC 2015 benchmark problems set and it was compared with a baseline of random immigrants strategy denoted by RND. The effects of change severity and change frequency on the performance of both algorithms were also studied. The experimental results have shown that MOEA/D-KF performs better than RND for type I DMOPs and presents competitive results on type II DMOPs

Table 3 Change prediction-based dynamic EAs

Algorithm	Compared to	Used benchmarks	Used performance metrics
D-QMOO [8]	–	FDA1 [17]	The objective error [17] and the design error [17]
The work of Hatzakis and Wallace [43]	–	FDA1 [17]	The objective error [17] and the design error [17]
DMEA/PRI [34]	–	FDA1 [17] and ZJZ [34]	The distance-based indicator [27] and HV Difference [34]
Dynamic MO-EGS [9]	dCCEA [44] and dPAES [45]	FDA1 [17], FDA3 [17], DIMP1 [9] and DIMP2 [9]	Variable Distance (VD) [10] and MS [10]
CDDMEA [37]	D-NSGA-II [6]	A test problem defined in [38] and FDA2 [17]	U-measure [46]
The work of Roy and Mehnen [36]	–	The gradient material problem [36]	–
RM-MEDA with PPS [40]	RM-MEDA with random initialization strategy and RM-MEDA with FPS	FDA1 [17], FDA4 [17], dMOP1 [10], dMOP2 [10] and F5-F8 [40]	IGD [32]
DMOEA/ADLM [41]	MM [6], VARM [8] and PREM [8]	FDA1 [17], FDA1E [41], FDA1L [41], FDA5 [17], FDA5E [41] and FDA5L [41]	The distance-based indicator [27]
MOEA/D-KF [42]	RND [42]	IEEE CEC 2015 Dynamic Benchmark Problems	IGD [32]

while RND performs marginally better on type III test problems. It was also observed that MOEA/D-KF faces many difficulties when solving problems with high change severity, isolated and deceptive fronts.

Since the prediction may not be always successful, there is a need to combine predictive-based approaches with a maintaining diversity mechanism. Moreover, these approaches are suitable only to dynamic environments presenting a behavior that follows a certain trend. Table 3 summarizes the algorithms discussed in this section.

4.3 Memory-Based Approaches

Memory-based approaches employ an extra memory that implicitly or explicitly stores the useful information from past generations to guide the future search. This technique is useful when optimal solutions repeatedly return to previous locations or when the environment slightly changes from one time step to another.

4.3.1 The Dynamic Competitive Cooperative CO-EA (dCOEA)

In [10], authors have presented a co-evolutionary multi-objective algorithm based on competitive and cooperative mechanisms to solve DMOPs. In order to overcome the difficulties of problem decomposition and subcomponent interdependencies arising in co-EAs, the proposed model addresses such an issue through emergent problem decomposition. In fact, the problem is decomposed into several subcomponents along the decision variables. These subcomponents are optimized by different species subpopulations through an iterative process of competition and cooperation. The optimization of each subcomponent is no longer restricted to one species but at each cycle, different subpopulations (i.e., competing species) solve a single component as a collective unit which permits the discovery of interdependencies among the species. The proposed competitive-cooperative CO-EA (COEA) is able to handle both static and dynamic multi-objective problems. In order to adapt COEA to DMOPs, authors have proposed to: (1) introduce diversity via stochastic competitors and (2) handle outdated archived solutions using an additional external population in addition to the archive. The proposed diversity scheme consists in starting the competitive mechanism, whenever a change is detected, independently of its fixed schedule in order to evaluate the adaptability of existing information within the various subpopulations with the new environment. Moreover, a set of stochastic competitors are introduced in addition to the competitors from the other subpopulations. If the winner is the stochastic competitor, the particular subpopulation is reinitialized in the region that the winner is sampled from. The external population denoted as the temporal memory is used in addition to the archive in order to store the potentially useful information about past PF since that the archived solutions will be discarded at the presence of an environmental change. The performance of COEA in static environments was tested against various multi-objective EAs (CCEA, NSGAII, and SPEA2) on different benchmark problems (FON, KUR, and DTLZ3). The obtained results have shown that COEA overcomes the others MOEAs in discovering near-optimal and well diversified PFs even for problems with severe parameter interdependencies. On the other hand, dCOEA was tested on four dynamic multi-objective test functions (FDA1 [17], dMOP1 [10], dMOP2 [10] and dMOP3 [10]) against two different dynamic MOEAs based on a basic MOEA and CCEA, respectively. The experiments were conducted at different change severity and frequency levels. The results have shown that dCOEA outperforms dMOEA and dCCEA in both aspects of tracking and finding a diverse PF. Nevertheless, the main drawback of dCOEA is its computational cost.

4.3.2 The Multi-strategy Ensemble MOEA (MS-MOEA)

Wang and Li [47] proposed new dynamic multi-objective test problems and a new Multi-strategy ensemble MOEA (MS-MOEA) where the convergence speed is accelerated using a new offspring generation mechanism based on adaptive genetic and differential operators. The proposed algorithm uses a Gaussian mutation operator and a memory-like strategy to handle population reinitialization when a change occurs. The basic process of the proposed algorithm is as follows. A set of sentry individuals are chosen randomly and their fitness values are re-evaluated. If the new values are different from the old ones, the population $P(t)$ and the archive $A(t)$ are re-initialized using the memory like strategy. In fact, the new generated populations are formed by two parts: (1) solutions randomly generated within the bounds of the search space and (2) solutions generated by the Gaussian local search operator. The proportion of these solutions is controlled by a probability p_l. Then, a number of parent solutions are selected from $P(t)$ and $A(t)$ in order to create only two offspring solutions c_1 and c_2. The proposed offspring creating strategy (i.e., GDM), uses simultaneously Genetic Algorithm (GA) and Differential Evolution (DE) in order to take advantages of both strategies. Finally, the population and the archive are updated by c_1 and c_2. The archive update is performed using the Fast Hypervolume (FH) strategy which consists in introducing the new solution in the archive only if it dominates an existing solution. This algorithm was compared against: (1) FH-MOEA, (2) MS-MOEADE which is similar to MS-MOEA but without the memory like re-initialization strategy and (3) the Improved NSGA-II (INSGA-II). INSGA-II is obtained by adding an archive population to maintain a set of non-dominated solutions found previously and by using a strategy of updating the archive that is an improved non-dominated selection based on crowding distances [47]. Two sets of experiments were conducted: experiments on static multi-objective problems and experiments on dynamic multi-objective problems including the FDA suite [17] and the proposed DMZDTs and WYL test problems [47]. The first set of experiments reveals the importance of cooperating the GDM strategy and the DE operators while the second set of experiments reveals the advantages of the multi-strategy ensemble. Nevertheless, the proposed approach is not suitable to problems with a low rate of change since it does not exploit any past information.

4.3.3 The Work of Wang and Li (2009)

Several memory-based dynamic environment handling schemes have been proposed in [48] to effectively reuse the useful past information to conduct the new population when the environment changes. These different schemes, including restart, explicit memory, local-search memory and hybrid memory schemes are based on the stored archive solutions. In fact, authors have proposed a DMOEA framework based on an improved version of the static MOEA NSGA-II [49]. The Improved NSGA-II algorithm, denoted as INSGA-II, is obtained by adding an archive population to maintain a set of non-dominated solutions found previously. The strategy of updating the archive is an improved non-dominated selection proposed in [27] and based on crowding distances. INSGA-II is used to conduct the selection, crossover, mutation and elite maintenance of the framework. Then, when a change occurs, the new population is composed by: (1) random solutions in addition to memory ones using the explicit memory scheme, (2) random solutions and solutions obtained by performing a Gaussian local search using the local-search memory scheme or (3) random solutions, memory solutions and solutions generated by application of a local search using the hybrid memory scheme. The comparative experiments were done using six dynamic multi-objective EAs conducted under the framework of dynamic INSGA-II by modifying the dynamic environment handling strategy and including the GA-DE strategy proposed in [47]. The test problems used were FDA1 [17], DMZDT1, DMZDT2, DMZDT3, DMZDT4, and WYL [47]. Two sets of experiments were conducted: (1) experiments on instances with small change rate and (2) experiments on instances with large change rate. The empirical results have shown that the proposed memory schemes improve the performance of the algorithm compared with restart scheme. Nevertheless, the higher the change degree is, the smaller the effectiveness of memory schemes is except the localsearch memory scheme which is much more robust since it puts less attention in past optimal solutions. Moreover, the hybrid memory scheme was not demonstrated to be efficient which can be explained by the fact that the merits of separate schemes are lost by their demerits.

4.3.4 The Adaptive Population Management-Based Dynamic NSGA-II (A-Dy-NSGA-II)

When the change degree is small, information gained from the previous run can be exploited and reused to accelerate the convergence speed. However, when changes are large, there is a small correlation between the optimal solutions after a change and those before the change. Thereby, random restart would be a suitable strategy. Based on this observation, Azzouz et al. [50] proposed an adaptive hybrid population management strategy using memory, Local Search (LS), and random strategies to effectively handle environment dynamicity for DMOPs. The proposed strategy is based on a new technique that measures the change severity, according to which, it adjusts the number of memory, LS, and random solutions to be used. Moreover, they proposed a dynamic version of NSGA-II, called Dy-NSGA-II, within which they

Table 4 Memory-based dynamic EAs

Algorithm	Compared to	Used benchmarks	Used performance metrics
d-COEA [10]	dMOEA [10] and dynamic CCEA [44]	FDA1 [17], dMOP1 [10], dMOP2 [10] and dMOP3 [10]	VD [10] and Maximum Spread (MS) [10]
MS-MOEA [47]	FH-MOEA [47], MS-MOEADE [47] and INSGA-II [48]	FDA1 [17], FDA2 [17], FDA3 [17], WYL [47] and DMZDT test functions [47]	IGD [32] and HV [51]
The work of Wang and Li [48]	–	FDA1 [17], WYL [47] and DMZDT test functions [47]	IGD [32]
A-Dy-NSGA-II [50]	The work of Wang and Li [48]	FDA1 [17], FDA2 [17], DMZDT test functions [47] and WYL [47]	IGD [32], HV ratio [31] and MS [10]

integrated the above mentioned strategies. The novelty of this work lies in combining several strategies while using them adaptively based on problem characteristics that are mainly: (1) the change frequency and (2) the change severity. The performance of the proposed strategies was assessed on the FDA benchmark suite [17] and DMZDT test problems [47]. It has been shown that the M-strategy-based Dy-NSGA-II (M-Dy-NSGA-II) needs to be accompanied by a diversity maintenance/introduction mechanism. The LS-strategy-based Dy-NSGA-II (LS-Dy-NSGA-II) gives a better performance due to its exploration aspect. Contrarily to memory strategies, the R-Strategy is useful when changes are large but it loses its effectiveness when changes are of a small degree. The AH-strategy-based Dy-NSGA-II (A-Dy-NSGA-II) is the only algorithm that was able to outperform most other algorithms in problems with both small and high change severities. When compared to memory-based algorithms proposed in the work of Wang and Li [48], A-Dy-NSGA-II algorithm outperformed all other algorithms on both instances with small change severity and with large one.

The main drawback of memory-based approaches is that memory is very dependent on diversity and should, thus, be used in combination with diversity-preserving techniques. Table 4 summarizes the algorithms discussed in this section.

4.4 Parallel Approaches

When dealing with DMOPs, the EA should be able to converge as fast as possible to the optimal PF before the next change appears. Parallel EAs are used in this context since they are considered as efficient algorithms with an execution time much less important than EAs with a sequential implementation. Parallel EAs use several subpopulations that evolve simultaneously on different processors while communicating some informations in a structured network [52]. EAs are very easy to parallelize. There is a variety of ways to implement parallel EAs such as the master-slave model, the independent runs model, the island model, cellular EAs, etc. [52].

4.4.1 The Dynamic Multi-objective Optimization EA (DMOEA)

In [53], Zheng proposed a Dynamic Multi-objective Optimization EA (DMOEA) where the population is divided into $m + 1$ multiple subpopulations where m is the number of objectives. Each subpopulation evolves according to one single objective using a cellular genetic algorithm while the last subpopulation optimizes the average value of all the objectives. The $m + 1$ subpopulations are supposed to converge to the extreme points of the PF and one point having the minimal average value of different objectives. Moreover, this algorithm utilizes hyper-mutation operator to deal with environment changes. In fact, when a change is detected, hyper-mutation is used to copy a certain number of elite solutions from the archive to the population, while the rest of the individuals are replaced by random individuals. To update the archive, DMOEA used a geometrical Pareto-selection algorithm. This approach sets an auxiliary point that is far away from the approximated PF. Then, each solution in the PF is lined to the auxiliary point, which permits to identify them by slopes. When inserting a new individual in the archive, it is compared only to the solutions that are located in the same slope region. The solution furthest away from the auxiliary point will be kept in the archive. DMOEA was evaluated on FDA1 [17], modified FDA2 [53] and modified FDA3 [53] with a change frequency equals to 2000 generations and on FDA4 [17] and FDA5 [17] with a change frequency equals to 5000 generations. The experimental results have shown that this algorithm is able to converge to changing PFs with well distributed points.

4.4.2 The Dynamic Version of Parallel Single Front Genetic Algorithm

Camara et al. [11] have proposed a procedure for the adaptation of the Parallel Single Front Genetic Algorithm (PSFGA) to dynamic environments. PSFGA is a parallel algorithm for multi-objective optimization that uses a master-worker architecture where the sequential algorithm is decomposed into several tasks that are run on different data distributed between several processors. PSFGA uses an island model where not only objective functions evaluations but also variation operators are concurrently done by every worker process. In fact, the population is divided into subpopulations of equal size distributed between different worker processes. On each subpopulation, the SFGA algorithm is executed for a fixed number of generations *genpar* and only the non dominated solutions are kept. Then, all workers send their affected sub-populations to the master process who joins all the solutions into a new population. Then, it runs an instance of the SFGA algorithm (along *genser* iterations) over the whole population. After performing a crowding mechanism for keeping the diversity, it sends new subpopulations again to the worker processes. This process is repeated until a stopping criteria is met. This algorithm has been evaluated on FDA1 and FDA2 test problems [17]. The empirical results have shown that the quality of the solutions worsens slightly as the number of workers used to solve the problem increases. Moreover, this approach is sensitive to the data decomposition which must be done on a balanced way to permit the speedup of convergence.

4.4.3 The Work of Camara et al. (2008)

In [54], a generic parallel procedure for dynamic problems using EAs was presented and used to compare the parallel processing of several multi-objective optimization EAs (i.e., SFGA, SFGA2, SPEA2, and NSGA-II). The proposed parallel procedure is based on an island model together with a master process that divides the population into several subpopulations of the same size to send to each worker process. Every worker uses the chosen multi-objective EA to search the optimal solutions in its subpopulation. After a fixed number of iterations (i.e., *genpar*), the workers send the non dominated solutions found to the master, who after grouping all the solutions into a new population, runs an instance of the same multi-objective EA (along *genser* iterations) over the whole population. Finally, the master sends again the new subpopulations to the worker processes. The different algorithms were evaluated using the FDA1 test function [17] in addition to proposed modified versions of the FDA2 and FDA3 test functions. It has been demonstrated the ability of the proposed procedure to reach PSs near to the optimal PSs in addition to the considerable reduction in the convergence speedup compared to the sequential algorithms.

Parallel approaches are effective methods to locate and track optimal PFs in dynamic environments. However, the main problem of these approaches consists in the difficulty of finding the most interesting decomposition. Table 5 summarizes the algorithms discussed in this section.

Table 5 Parallel dynamic EAs

Algorithm	Compared to	Used benchmarks	Used performance metrics
DMOEA [53]	–	FDA1 [17], modified FDA2 [53], modified FDA3 [53], FDA4 [17] and FDA5 [17]	Running time and HV [51] and the size of non-dominated set
Dynamic PSFGA [11]	–	FDA1 [17] and FDA2 [17]	HV [51], accuracy [11] and stability [11]
The work of Camara et al. [54]	–	FDA1 [17], modified FDA2 [54] and modified FDA3 [54]	HV [51], the execution time and the size of non-dominated set

4.5 Approaches that Convert the DMOP into Multiple Static MOPs

4.5.1 The Work of Wang and Dang (2008)

When the environment changes are gradual and continuous, it is very difficult to an optimization algorithm to rapidly react to changes and to continually converge to optimal solutions relatively to each change. This is why, Wang and Dang [55] proposed to obtain Pareto optimal solutions at some representative time instants instead of low quality solutions at all the time. To do so, they proposed to convert the DMOP into multiple static MOPs by dividing the time period of the DMOP into several smaller time intervals. For each time interval, the original DMOP is seen as a static Multi-objective Optimization Problem (MOP) with objective functions and constraints remaining unchanged over time. Thus the DMOP is approximated by a series of static MOPs. Moreover, each static MOP is transformed into a bi-objective optimization problem. The first objective is related to population diversity and the distribution of solutions using a defined U-measure. The second objective is to increase the quality of the found non-dominated solutions using a non-domination ranking. A new uniform crossover operator is used to avoid crossover between parents that are too close to each other during the beginning of the algorithm run. As well, a new selection scheme is proposed to find Pareto optimal solutions in different regions and for the different time periods. The proposed algorithm was evaluated on FDA1, FDA2 and FDA3 test problems [17] and it was confronted to static NSGA-II [49]. The experimental results have shown that the proposed EA is able to effectively track time changing PFs and it has a better performance than NSGA-II with respect to the coverage metric and the uniformity measure.

4.5.2 The Dynamic Multi-objective EA (DMEA)

Liu and Wang [56] presented a new dynamic EA called DMEA where the time period of the DMOP is divided into multiple smaller equal subperiods where each one is seen as a fixed environment. In each subperiod, the DMOP is optimized as a static MOP using an EA. The same as in [55], the static MOP is converted into a bi-objective optimization problem with one objective is the static rank variance and the second one is the density variance. Moreover, a new environment changing feedback operator is defined to check out environment variations. The performance of DMEA was evaluated only on two DMOPs: (1) G1 test function which was proposed in this work and (2) G2 which was developed in [38]. Only PF plots were presented and no performance measures were used but authors noted that according to the presented plots, the algorithm was able to track changing PFs. DMEA was more evaluated in [57] on four test problems which are G1 [56], G2 [38], G3 (i.e., FDA2 [17]) and G4 (i.e., FDA3 [17]). No performance measures were used in this study as well and only plots of the obtained PFs were presented. Helbig et al. [56] noted that although the

Table 6 Dynamic EAs that convert the DMOP into multiple static MOPs

Algorithm	Compared to	Used benchmarks	Used performance metrics
The work of Wang and Dang [55]	NSGA-II [49]	FDA1 [17], FDA2 [17] and FDA3 [17]	C-metric [59] and U-measure [46]
DMEA [56]	–	G1 [56] and G2 [38]	–
DMEA [57]	–	G1 [56], G2 [38], FDA2 [17] and FDA3 [17]	–
DSG [58]	DEG [60] and DFA [61]	DMT1-DMT4 [58]	C-metric [59] and HV [51]

authors of DMEA claimed that, with respect to the presented plots, their algorithm is able to converge to optimal PFs, this is not the case. Helbig et al. noted that the algorithm lost track of the changing PF for FDA2 test problem.

The same idea of DMEA was borrowed in [58] to be adapted to constrained optimization where a new fitness selection operator was proposed. It permits to select individuals that will participate in the next generation according to the number of feasible solutions in the population. If this number is greater than the maximum population size, infeasible solutions are discarded and only feasible one are ranked based on a dynamic mean rank variance. Otherwise, feasible solutions are maintained and the rest of the population is formed by infeasible solutions ranked based on their density. Although this algorithm called DSG, is developed to handle constrained DMOPs, it was evaluated on unconstrained test functions which are extensions of FDA1 [17], FDA2 [17], FDA3 [17] and a test function proposed in [38]. Table 6 summarizes the algorithms discussed in this section.

4.6 Other Approaches

4.6.1 The ALife-Inspired Algorithm for DMOPs

Amato and Farina [62] have proposed an artificial life-inspired EA for dynamic multi-objective optimization in the case of unpredictable parameters changes. Contrarily to classical EAs where the Darwinian evolution is considered as a type of intelligence, the proposed method considers that life and interactions among individuals in a population in a changing environment is itself a type of intelligence to be exploited. The proposed algorithm considers the coded strings as individuals interacting in a population rather than simple individuals genotypes. Thereby, the artificial operators imitate interactions between individuals such as meeting, fight and reproduction [62]. It is noting that in this approach there is not a selection operator. Then, all individuals have a similar probability to survive. At each generation, an individual is considered. He can meet or not another individual according to a probability p_m. When meeting occurs, either bisexual reproduction or competition (*fight*) may take place. If bisexual reproduction has occurred, two new individuals are then added to the population. Otherwise, fight is performed between the two selected individuals. In this case, the objective functions are evaluated for both individuals. Then, only the Pareto dominating individual survives. If nobody dominates the other, the individual in the more crowded region is eliminated in order to preserve diversity. Then, the population size is reduced by one. If meeting does not occur, asexual reproduction may be performed with probability p_{ar} equal to p_{br}, which adds a new individual to the population. Authors have noted that the proposed algorithm is supposed to run for an indeterminate time following environment change, without definitely converging towards a final optimum unless a static system is considered. For test problems a fictitious maximum iteration or generation number is imposed [62]. The proposed algorithm was tested on the FDA1 test problem [17]. The results have shown that the algorithm converges slowly especially after a sudden change where the convergence to the new optimal set was much more slower than the previous one. This was explained by the absence of a fitness based selection.

4.6.2 The Dynamic Orthogonal Multi-objective EA (DOMOEA)

In [63] authors developed a Dynamic Orthogonal MOEA called DOMOEA, which presents a generalization of the Orthogonal MOEA (i.e., OMOEA-II) to dynamic environments. It deals with problems having continuous decision variables, where the objective functions change with time while the number of objective functions and the number of decision variables are static. The process of the proposed algorithm is as follows. After the population initialization, the crossover operator is performed on the population P_t giving rise to the population of offspring solutions Q_t with the cardinality N_p. Two types of crossover operations are used: (1) the orthogonal crossover executed with the probability p and (2) the linear crossover executed with the probability $1 - p$. After the crossover operation, P_t and Q_t are combined in the population R_t, on which the selection operator is performed in order to get the next population P_{t+1}. This operator is based on the sorting method used on NSGA-II and the clustering technique of SPEA2 to maintain diversity. Finally, if an environmental change has been detected, P_t is defined as the current approximated optimal PS and all parameters are reinitialized; otherwise, the above described process is repeated. The proposed algorithm was tested on the FDA test suite [17]. However, only the results of the first three dynamic problems with two objectives were presented. The obtained results have shown the ability of the algorithm to track and find a diverse PF. One of the disadvantages of this approach is that the statistical method used (i.e., the orthogonal design method) has been proven to be optimal for only additive and quadratic models. Moreover, since DOMOEA uses the current population, as an initial population when a change is detected, it may be sensitive to problems with high change's degree. Thus, the performance of the proposed approach has to be tested with different environmental change severities.

4.6.3 The Work of Deb (2011)

More recently, Deb [64] presented two different approaches that are usually used when resolving dynamic single-objective as well as multi-objective optimization problems. The first approach consists in developing a set of optimal knowledge base to be used as guiding rules for handling changing problems on-line. This approach is useful for problems with frequent changes and it is computationally expensive for any optimization algorithm to be applied on-line. The second one is an on-line optimization approach in which an off-line study is used to find a minimal time window within which the problem will be considered and treated as a static problem. This approach is more appropriate for slow changing problems. Moreover, an automated decision-making approach based on the utility function concept has been proposed since a solution should be chosen and implemented as quickly as the PF is found, and before the next change appears. An utility function was used to provide different weights to different objectives. Then, the chosen solution is the middle point in the trade-off frontier providing a solution equidistant from individual optimal solutions. The first approach was applied to a robot navigation problem which consists in finding an obstacle-free path which takes a robot from a point A to a point B with minimum time. Since the imprecise definition of the deviation in this problem, a genetic-fuzzy approach was proposed based on a genetic algorithm which is used to create the knowledge base composed of fuzzy rules for navigating a robot off-line. Then, for on-line application, the robot uses its optimal knowledge base to find an obstacle-free path relatively to a given input of parameters that represents the state of moving obstacles and the state of the robot. The second approach was applied to a bi-objective hydro-thermal power scheduling problem using a previously proposed modified NSGA-II procedure which has identified a minimum time window of 30 min in which the power demand can be considered stationary.

4.6.4 The Dynamic Multi-objective Optimization Algorithm Inspired by P Systems (DMOAP)

In [65], authors designed several special test functions in addition to a dynamic MOEA inspired by P systems called DMOAP. This latter is based on membrane computing where the global system is composed of $m + 1$ subsystems: m subsystems are single-objective optimization subsystems that only optimize a corresponding objective while an additional subsystem is relative to the true multi-objective optimization subsystem that optimizes all objectives simultaneously. Each subsystem contains several membranes. The membrane has its own subpopulation and works like a single EA. These membranes are contained within two special membranes that collect the resulting chromosomes from subsystems and in which the chromosomes will not evolve. Furthermore, in this paper DMOPs were classified into two types: slow-change problems and fast change problems. Slow change problems are characterized by a long static state. Thus, the dynamic problem can be divided into n Static MOPs (SMOPs) and the optimal PS of the DMOP can be approximated by the superimposition of the optimal solutions of each SMOP on different instants. However, if the time period needed by the EA to improve its candidate solutions is more important than the time period during which the objectives are assumed to be stationary, the problem is considered to be a fast-change problem that will be transformed to a slow-change problem. This transformation concerns the objective functions. The proposed membrane control strategy has been applied to the optimal control of a time-varying unstable plant that has been presented as a dynamic multi-objective optimization problem. Simulation results demonstrated that the proposed strategy has an excellent performance in terms of stability, real-time performance and reliability although the proposed model is executed on a serial computer. The best model is that all membranes evolve in parallel [65].

4.6.5 The Multiple Reference Point-Based MOEA (MRP-MOEA)

Multiple Reference Point-based MOEA (MRP-MOEA) [66] deals with dynamic problems with undetectable changes. This algorithm does not need to detect changes. It uses a new reference point-based dominance relation ensuring the guidance of the search towards the optimal PF. The main idea behind MRP-MOEA is to define multiple targeted search directions (also known as goals) and to seek simultaneously the location of the optimal solutions along these different directions, rather than searching in the whole search space. Since several optimal points can be found relatively to different Reference Points (RPs) generated in a structured manner and covering the entire search space, the algorithm may be able to converge quickly to the desired PF without needing to detect changes. To generate this set of uniformly distributed RPs, authors used Das and Dennis's method. It generates K points on a normalized hyperplane with a uniform spacing δ in each axis, for any number of objectives M. The framework of the proposed algorithm is based on NSGA-II with significant changes in the non-domination sort mechanism and some other extensions such as the use of a LS technique at the beginning of each generation. The goal of the LS is to ameliorate existing solutions and to detect the new search directions whenever a change appears. Moreover, in order to provide well-distributed solutions along the PF, an archive update strategy was designed to maintain representatives of all prominent RPs. The proposed algorithm was tested on the FDA test suite [17] and the dMOP test problems [10]. Simulation results have shown that MRP-MOEA permits not only to track the PF but also to maintain diversity over time albeit the changes are undetectable. The algorithms discussed in this section are summarized in Table 7.

Table 7 Non classified dynamic EAs

Algorithm	Compared to	Used Benchmarks	Used Performance metrics
The ALife inspired algorithm [62]	–	FDA1 [17]	–
DOMOEA [63]	–	FDA1 [17], FDA2 [17] and FDA3 [17]	GD [33] and Spread [49]
The work of Deb [64]	–	Robot navigation problem and hydro-thermal power scheduling problem	HV ratio [31]
DMOAP [65]	–	Optimal control of a time-varying unstable plant problem	–
MRP-MOEA [66]	d-COEA [10], dCCEA [44] and dMOEA [10]	FDA1 [17], dMOP1 and dMOP2	VD [10], IGD [32], HV ratio [31] and MS [10]

5 Test Functions for Dynamic Multi-objective Optimization

5.1 Synthetic Test Functions

Benchmark test problems are functions with specific challenging characteristics that permit to evaluate the ability of an algorithm to solve DMOPs and to efficiently overcome different difficulties that can occur in real-world problems.

In [38], Jin and Sendhoff proposed an approach for constructing dynamic multi-objective test problems by aggregating objective functions of existing stationary test problems through dynamically changing weights. This approach has been used by several other researchers [56, 67, 68].

Farina et al. have proposed in [17] the first suite of dynamic multi-objective test problems, called FDA benchmark functions, by adding time-varying terms to the objectives in stationary multi-objective test problems (ZDT and DTLZ). The FDA test functions are of type I, II and III while the number of decision variables, the number of objectives and constraints boundaries keep fixed. Also, the optimal PF may be convex, concave or changing from convex to concave over time. One of the advantages of the FDA functions is that they are easy to construct, and the number of decision variables are easily scalable [39]. Therefore as noted in [39], the FDA test suite exhibits the characteristics, defined by Deb [28], that benchmark functions should have. This is why, this test suite was used by several researchers who developed different extensions of these functions. A generalization of the FDA test functions was proposed in 2006 [33] where several parameters such as the number of disconnected optimal PFs and the spread of solutions can be simply specified. Similar to the FDA test suite [17], Tang et al. [69] also proposed to construct dynamic test functions on the basis of the ZDT functions [28]. Moreover, they presents an additional explanation of how to calculate the POF. In 2007, Zhou et al. [34] proposed a modified version of FDA1 where they incorporated nonlinear dependencies between the decision variables. The modified FDA1 function is called ZJZ. ZJZ is a Type III test problem. As well, in 2009, Goh and Tan [10] have proposed three dynamic multi-objective test problems called dMOP1, dMOP2 and dMOP3 based on the FDA ones. dMOP1 is a Type III test problem while dMOP2 is a Type II one and they both have a POF that changes from convex to concave over time. In contrast to the FDA2 problem where the POF changes from a convex to a concave shape only for specific values of the decision variables, dMOP1 and dMOP2 have the advantage of not being sensitive to this problem.

In 2005, Guan et al. have proposed to create dynamic multi-objective test functions by replacing some objectives with new objectives during evolution [60]. In this approach, the objective functions should be selected carefully in order to permit to evaluate the performance of EAs in different ways. Avdagic et al. [70] proposed an adaptation of the DTLZ problems to dynamic environments. They developed the following types of test functions: (1) type I DMOP where the POS changes coherently over time but the POF remains invariant; (2) type II DMOP where the shape of the POS continuously changes and the POF changes over time; and (3) type II DMOP where the number of objective functions changes over time [70]. Koo et al. have proposed two new benchmark functions called DIMP1 and DIMP2 in 2010 [9] where unlike FDA and dMOP test problems, each decision variable has its own rate of change. Wang and Li have also proposed new type I DMOPs based on the ZDT functions [47]. Motivated by the observation that all previous dynamic multi-objective test problems assume that the current optimal PS or optimal PF does not affect the future one, Huang et al. have proposed four dynamic multi-objective test problems called T1, T2, T3 and T4 in [65]. Since the FDA and dMOP suites contain only DMOPs with continuous optimal PFs, Helbig and Engelbrecht [71] developed two DMOPs named HE1 and HE2 that are based on the ZDT3 test function with a discontinuous POF. Recently, they proposed in [72] three new dynamic multi-objective test functions with complex POSs where the POS is different for each decision variable. In 2014, a comprehensive overview of existing dynamic multi-objective benchmark functions was provided in [39] while highlighting their shortcomings. Moreover, to address the identified problems, authors proposed new benchmark functions with complicated POSs, and approaches to develop DMOPs with either an isolated or deceptive POF. As well, Biswas et al. [73] proposed some general techniques to design DMOPs with dynamic PS and PF through shifting, shape variation, slope variation, curvature variation, etc. They proposed 9 benchmark functions derived from the benchmark suite used for the 2009 IEEE Congress on Evolutionary Computation competition on static bound-constrained multi-objective optimization algorithms. These test functions are denoted as UDF1-UDF9.

Although there is a number of dynamic multi-objective test functions that were proposed, there is a lack of those taking into account simultaneously time-dependent objective functions and constraints. In 2015, Azzouz et al. [7] proposed a set of benchmark functions, called Dynamic CTPs (DCTPs), that extend the CTP suite of static constrained MOPs where the PF, the PS and the constraints are simultaneously time-dependent. These characteristics make the task of optimization much more difficult than dynamic unconstrained problems. In addition, these test functions present two kinds of tunable difficulties in a multi-objective optimization EA: (1) difficulty in the vicinity of the optimal PF where constraints do not make a major portion of the search space infeasible except near the optimal PF (the case of DCTP1 to DCTP5), and (2) difficulty in the entire search space where constraints produce different disconnected regions of feasible objective space (the case of DCTP6 to DCTP8).

5.2 Real-World Applications

Several real-world dynamic multi-objective optimization applications exist in the literature. Helbig and Engelbreght [14] grouped and classify the main important areas of these applications as follows:

- **Control problems**: including the controller design for a time-varying unstable plant [17, 65], the regulation of a lake-river system [74], the optimization of indoor heating [75], and the control of a greenhouse system for crops [76].
- **Scheduling problems**: such as the hydro-thermal power scheduling problem [6], and the job-shop scheduling problem [77, 78].
- **Resource management problems**: such as war resource allocation optimization [79] and the management of hospital resources [80].
- **Routing problems**: several real world applications belong to this category such as route optimization according to real-time traffic [81], the routing problem in mobile ad hoc networks [82], the dynamic vehicle routing problem [83, 84], the robot navigation problem [64] and the optimization of supply chain networks [85, 86].
- **Mechanical design problems**: such as the machining of gradient material [36] and design optimization of wind turbine structures [87].

Table 8 presents a summary of the most used dynamic test functions and real world problems and their references.

6 Performance Assessment of Dynamic MOEAs

6.1 Performance Metrics

When solving an optimization problem, there is a need to assess and measure the performance of different algorithms and to evaluate the quality of their obtained solutions. This is to compare and rank their effectiveness with respect to different

Table 8 Table of most used dynamic test functions and real world problems

Category	Problem	Referenced in
Synthetic problems	FDA test suite [17]	[6, 8–11, 17, 34, 37, 40, 41, 43, 47, 48, 50, 53–55, 57, 62, 63, 66]
	Three problems proposed in [60]	[60]
	DSW suite and DTF [33]	[33]
	dMOP test suite [10]	[10, 40, 66]
	DIMP1 and DIMP2 [9]	[9]
	DMZDT test suite and WYL [47]	[47, 48, 50]
	T1, T2, T3 and T4 [65]	[65]
	Four test problems proposed in [40]	[40]
	DCTP test suite [7]	[7]
Real world problems	Control problems	[17, 65, 74–76]
	Scheduling problems	[6, 64]
	Routing problems	[64, 81, 82, 85, 86]
	Resource management problems	[79, 80]
	Mechanical design problems	[36, 87]

requirements such as convergence, diversity, spread of solutions, etc. This is why, the choice of appropriate measures and statistical tests is very important to produce a fair comparison.

When dealing with static problems it is generally often enough to just evaluate the final population that the algorithm converges to at the end of the search process. However, in a dynamic context the performance metrics should not only assess the quality of the final population but also evaluate the robustness of the resolution algorithm facing changing environments. This includes how well the algorithm is able to detect problem changes and to discover the new promoting search areas and to track optimal solutions as they move in the search space. Using just the population quality at one time point is not fair enough since it may be possible that one algorithm has a good population at one time step but it loses optimal solutions in the rest of the optimization process while another algorithm has a worser final population but it have keeped tracking optimal solutions all over changing environments.

Several performance metrics were proposed in the literature to evaluate the performance of dynamic multi-objective optimization algorithms. In the following, we will survey the most commonly used ones.

6.1.1 Accuracy Performance Measures

- The Generational Distance measure (GD): The Generational Distance (GD) is a metric developed for stationary multiobjective optimization which measures the distance between the optimized optimal PF and the true one. In [33], Menhen et al. have proposed to calculate the GD metric in the decision space since some DMOPs have optimal PSs that dynamically change over time. The new metric called G_τ approximates the distance between the current optimal PS and the true one. Goh and Tan [10] also adopted the calculation of the GD metric in the decision space. The proposed performance measure, named the variable space generational distance metric (VD), measures the closeness of the approximated PF to the optimal one. The VD metric is calculated as follows:

$$VD_{offline} = \frac{1}{\tau} \sum_{t=1}^{\tau} VD * I(t) \tag{2}$$

$$VD = \frac{\sqrt{\mid PF \mid \sum_{v \in PF} d(v, PF^*)^2}}{\mid PF \mid} \tag{3}$$

$$I(t) = \begin{cases} 1, & if\ (t\%\tau_T) = 0 \\ 0, & otherwise \end{cases} \tag{4}$$

where t is the current iteration number, τ_T is the change frequency, % is the modulus operator, PF is the obtained PF and PF^* is the true optimal PF.

Several other works have been proposed in this topic such as the $rGD(t)$ metric proposed in [67].

- The Inverted Generational Distance metric (IGD): The IGD metric proposed by Sierra and Coello [32] gives an indication of the distance between the optimal PF and the evolved PF. In addition to the convergence, the IGD can measure the diversity of the obtained PF. Mathematically it is defined as follows:

$$IGD(PF, PF^*) = \frac{\sum_{v \in PF^*} d(v, PF)}{|PF^*|} \tag{5}$$

where PF is the obtained PF, P^* is a set of uniformly distributed points along the optimal PF in the objective space and $d(v, PF)$ is the minimum Euclidean distance between v and the points in PF. The smaller the IGD value is, the closer PF is to the optimal PF. In [48], Wang and Li proposed to use the mean IGD metric calculated as follows:

$$\overline{IGD} = \frac{1}{nbChanges} \sum_{i=1}^{nbChanges} IGD_i \tag{6}$$

where $nbChanges$ is the number of occurred changes and IGD_i is the IGD value calculated before the occurrence of the $(i+1)$th change.

- The Success Ratio: The success ratio proposed in [33] indicates the ratio of the found solutions that are members of the true optimal PF and is defined as follows:

$$SC = \frac{|\{x \backslash f(x) \in PF^*\}|}{|PF|} \tag{7}$$

where PF^* and PF are respectively the true optimal PF and the current one. The main drawback of this metric is that if an algorithm obtains a high number of solutions not Pareto optimal but very close to the optimal PS, it will have a success ratio inferior than one algorithm having only one solution belonging to the true optimal PS.

6.1.2 Diversity Performance Measures

- The maximum spread: The adaptation of the maximum spread metric to dynamic multi-objective optimization (MS') was introduced in [10] and is defined as follows:

$$MS'(PF, PF^*) = \sqrt{\frac{\sum_{j=1}^{M}\left(\frac{min(PF_{j,u}, PF^*_{j,u}) - max(PF_{j,l}, PF^*_{j,l})}{PF^*_{j,u} - PF^*_{j,l}}\right)^2}{M}} \tag{8}$$

where $PF_{j,u}$ and $PF_{j,l}$ are respectively the maximum and the minimum value of the j-th objective in the obtained PF. $PF^*_{j,u}$ and $PF^*_{j,l}$ are respectively the maximum and the minimum value of the j-th objective in the optimal PF. MS' is applied to measure how well the optimal PF is covered by the obtained PF. A higher value of MS' reflects that a larger area of PF^* is covered by PF.

- The Path Length measure (PL): Since most of the proposed diversity measures use the Euclidan distance, they do not take into account the shape of the PF. Thus, a new measure based on path length for calculating distance between solutions is proposed in [33]. The PL measure is the normalized product of the path between sorted neighbouring solutions on the optimal PF.
- The Set Coverage Scope (CS): The Coverage Scope (CS) measure was introduced by Zhang and Qian in [88]. It quantifies the coverage of the non-dominated set by averaging the maximum distance between each solution and the other solutions in the obtained PF. CS is calculated as follows:

$$CS = \frac{1}{|PF|}\sum_{i=1}^{|PF|} max\{\| f(x_i) - f(x_j) \|\} \tag{9}$$

where PF is the obtained optimal PF and $x_i,\ x_j \in PF$ with $i \geq 1$ and $j \leq |PF|$.

6.1.3 Robustness Performance Measures

- The Stability measure: The stability measures the effect of environment changes on the accuracy (i.e., acc) of the algorithm. It was firstly proposed for dynamic single-objective optimization in [89] and it was adapted for dynamic multi-objective optimization in [90]. This measure is defined as follows

$$stb(t) = \begin{cases} stb_0(t) & if\ stb_0(t) \geq 0 \\ 1 & otherwise \end{cases} \tag{10}$$

$$stb_0(t) = acc(t) - acc(t-1)$$

- The Reactivity measure: This metric measures the ability of an algorithm to react to changes by evaluating how much time the algorithm takes to achieve a desired accuracy threshold. Similar to the stability, the reactivity measure is an adaptation of a previous version developed by Weicker in [89] for dynamic single-objective optimization. This measure was adapted for dynamic multi-objective optimization in [90] and is defined in the following

$$react_{alternative,\epsilon}(t) = min\{\{t' - t \mid t < t' \leq maxgen,\ t' \in N,\ acc(t') - acc(t) \geq \epsilon\} \cup \{maxgen - t\}\}$$

where $maxgen$ is the maximum number of generations.

6.1.4 Combined Performance Measures

This kind of measures are used to take into account several aspects simultaneously in order to evaluate the overall quality of the obtained optimal PF.

- The Accuracy measure: The accuracy measures the closeness of the current best found PF to the true optimal PF. Camara et al. [11] proposed to calculate the accuracy based on the ratio of the hypervolume of the current approximated PF and the maximum hypervolume ($HVmax$) that has been found so far. The accuracy is calculated as follows:

$$acc_{maximization}(t) = \frac{HV_{max}}{HV(PF(t))} \tag{11}$$

$$acc_{minimization}(t) = \frac{HV(PF(t))}{HV_{max}} \tag{12}$$

$$acc(t) = \begin{cases} acc_{maximization} & if\ objectives\ are \\ & maximized \\ acc_{minimization} & if\ objectives\ are \\ & minimized \end{cases} \quad (13)$$

- The Hypervolume difference: Zhou et al. [34] proposed to use the hypervolume difference (HVD) to evaluate the quality of the found optimal PF. HVD is calculated as follows:

$$HVD = HV(PF^*) - HV(PF) \quad (14)$$

The problem with this metric is that it can not be used when the true optimal PF is unknown. In the same context, Camara et al. [90] extended the definition of the accuracy measure for the case when the true optimal PF is known. The new accuracy, noted as acc_{alt} is defined as the absolute value of the HVD at time t and is calculated as follows:

$$acc_{alt} = |HV(PF^*) - HV(PF)| \quad (15)$$

- The hypervolume ratio: The hypervolume of a set A with respect to a reference point ref noted as $HV(A, ref)$ is the hyperarea of the set $R(A, ref)$. $HV(A, ref)$ measures how much of the objective space is dominated by A [51]. The hypervolume ratio defined in [31], is calculated as follows:

$$HVRatio(PF, ref) = \frac{HV(PF, ref)}{HV(PF^*, ref)} \quad (16)$$

where PF^* is a set of uniformly distributed points along the true optimal PF in the objective space. The maximum value of the $HVRatio$ is 1 and as it becomes smaller, the performance of the algorithm is worser. Table 9 presents a summary of the most used performance metrics in dynamic multi-objective optimization.

6.2 Comparing the Performance of Different Algorithms

Given a set of algorithms and their performance evaluation values, comparing and ranking these various algorithms is not a trivial task. Several works in the literature simply runned several instances of the algorithm. Then, they calculated, for each performance measure the average and the standard deviation. The algorithms are then ranked based on these values [14]. It should be noted that typically various performance metrics are used. One algorithm may perform very well with respect to some measures while it may not be the case regarding some others. This is why, ranking different algorithms should be performed with respect to each performance metric separately. Moreover, the use of statistical tests instead of simply comparing the mean and standard deviations values becomes more and more essential. When

Table 9 The most used performance metrics in dynamic multi-objective optimization

Category	Performance metric	Referenced in
Accuracy measures	GD [33]	[30, 62]
	VD [10]	[9, 10, 66]
	IGD [32]	[7, 40, 42, 47, 48, 50, 66]
	SC [33]	–
Diversity measures	MS [10]	[7, 9, 10, 50, 66]
	PL [33]	–
	CS [88]	–
Robustness measures	Stability measure [89, 90]	[11, 90]
	Reactivity measure [90]	[90]
Combined measures	Accuracy measure [11, 89, 90]	[11, 90]
	HV [51]	[11, 47, 53, 54, 58]
	HVD [34]	[34]
	HV ratio [31]	[6, 7, 50, 64, 66]

Table 10 The most used statistical tests in dynamic multi-objective optimization

Type	Statistical test	Referenced in
Parametric	t-test	[40, 47]
Non-parametric	Kolmogorov–Smirnov test	[9, 10]
	Wilcoxon test	[7, 50, 66]

analyzing the literature, we observed that several works just reported the mean and deviation values while some others used parametric statistical tests like Student's t-test. Here, we note that the use of such tests should be preceded by the verification that the performance values follow a normal distribution. This is why, the use of non-parametric statistical tests such as the Wilocoxon test becomes more and more considered by different authors. It confirms that the difference between two populations of values (performance metrics values) is not obtained by chance. Table 10 presents the most used statistical tests in dynamic multi-objective optimization.

7 Discussion

Recently, a number of population-based approaches, including EAs, artificial immune systems and particles swarm optimization approaches have been proposed and applied to solve DMOPs. Nevertheless, many challenges still not being taken into consideration.

7.1 General Challenges for Dynamic Optimization

When analyzing the literature of this research field, we remarked that there is a lack of standardisation. First of all, there is no standard dynamic multi-objective benchmark functions. For this reason, the performance of the proposed dynamic algorithms were evaluated differently using different test functions. The same observation is made concerning the performance metrics. Thus, it is difficult to fairly compare the existing works unless re-implementing all of them and re-evaluating their performance. Moreover, statistical tests are not yet highly used although their importance and their usefulness to produce a fair comparison between different approaches. Studies presenting a comprehensive state of the art of existing benchmark functions and existing performance measures are very required. As well, a statistical comparative study of representative works of different approaches and using standard test functions and performance metrics is needed. This is to understand their behaviours facing different challenging types of DMOPs.

7.2 Specific Challenges for Dynamic MOEAs

This chapter was mainly devoted to provide a survey of the research that has been done over the past decade on the use of specially EAs for dynamic multi-objective optimization. Concerning this specific research topic, in addition to the above mentioned general challenges, we have observed a lack of works on mainly three directions:

- **Dynamic constrained optimization**: In real world, we often encounter problems that not only involve the optimization of several conflicting objectives simultaneously but also have a set of constraint conditions that must be satisfied. Several constraint handling techniques have been developed to be incorporated into EAs. Most of them are restricted to the static optimization. Despite the growing interest given to the use of EAs to solve dynamic optimization problems, most of the research was focused on the unconstrained or domain constrained problems. Applying EAs to solve constrained DMOPs is not yet highly explored although this kind of problems is of significant importance in practice. Many real-world problems are constrained DMOPs such as optimal control problems, portfolio investment, chemical engineering design like the dynamic hydro-thermal power scheduling problem, dynamic scheduling and transportation problems such as the dynamic multi-objective vehicle routing problems and so forth. In fact, when dealing with such problems, the main difficulties consist on the need to not only efficiently handle the constraints but also rapidly and continually track the changing PF and drive infeasible solutions to feasible ones whenever the constraints change. As presented in Sect. 4, very few studies are available in this direction [6, 7]. As well, we have observed a lack of benchmarks that simultaneously take into account the dynamicity of objective functions and constraints. Recently, Azzouz et al. [7] proposed the Dynamic CTPs (DCTPs) test functions, that extend a suite of static constrained MOPs where the PF, the PS and the constraints are simultaneously time-dependent. More studies in this research direction are required.

- **Dynamic parallel approaches**: When dealing with DMOPs, a time restriction is imposed since the EA should be able to converge as fast as possible to the optimal PF before the next change appears. Parallel EAs are used in this context since they are considered as efficient algorithms with an execution time much less important than EAs with a sequential implementation. Despite this interesting feature, regarding the works proposed in the literature, the use of parallel approaches represents the least focused research direction [11, 53, 54]. Investigating more efforts in developing such approaches would be very promoting.
- **Automatic Decision making**: When the decision maker has specific preferences, the EA should be able to converge the search towards the region of interest of the optimal PF. Such goal was highly studied in static environments in both cases of single and multiple decision makers [91–94]. However, a dynamic context might suggest the user preferences change over time and so the preference handling technique should allow preferences to be interactively adapted or automatically learnt during the optimization process. To the best of our knowledge, only few works [6, 77] proposed to suggest a decision-making aid to help identify one dynamic single optimal solution. This research direction is not yet highly explored.

8 Conclusion and Future Research Paths

In addition to the challenge of satisfying several competing objectives, industrial problems and many other problems that occur in our daily life are also dynamic in nature. In such a situation, the objective functions, constraints and/or problem parameters may change over time. Despite of the considerable number of approaches developed on dynamic single-objective optimization, dynamic multi-objective optimization is explored only recently. Several works have been established in the literature such as diversity-based approaches, change prediction-based approaches, memory-based approaches, parallel approaches, approaches that convert the DMOP into multiple static MOPs, etc. The objective of this chapter was to provide an overview of existing EAs proposed for the resolution of DMOPs. Moreover, a review of the most commonly used benchmark functions, real-world DMOPs, performance measures and statistical tests was presented. Challenges and future research directions were also discussed. This review has shown that several EAs have already been developed to solve DMOPs. Despite of all existing works, there still exist a need to future research in this area as the number of real world problems belonging to this category is in a dramatic increase. We have presented in Sect. 5.2 a summary of those that have been studied in the literature. However, due to the continuous increase of senior people and greater need for health, disability support and higher quality of life in general, some new real world problems such as smart houses and smart cities problems begun to be considered as important topics. We have focused in this problematic in [95] where we have modeled appliances scheduling as a dynamic constrained multi-objective optimization problem and have used DC-NSGA-II [7] for the problem resolution. Moreover, as generally, there are multiple inhabitants in the same home sharing context-aware applications with various conflicting individual preferences, we proposed a new comfort function to support multi-user conflictual preferences. The application of population-based approaches to smart houses and smart cities problems has not been highly studied. In this context, we suggest two main future research directions:

1. As smart technologies are considered as viable solution to maintain independence, functionality, well-being and higher quality of life, this motivate more research on this topic. Exploring the eligibility of dynamic EAs to solve problems revealed by smart houses and smart cities technologies may be of a significant importance. As well, the use and the evaluation of the performance of different population-based metaheuristics such as artificial immune systems [3, 88, 96], particles swarm optimization [14, 67, 71] and chemical reaction optimization [97] to solve such problems would be appreciated.
2. Smart houses and smart cities problems are strongly dependent on user preferences. A dynamic context might impose taking into account the change of these preferences over time and relatively to environment changes. The resolution method should be able to automatically learn decision maker's preferences during the optimization process. Such decision-making aid help identify the more interesting solutions or even one dynamic single optimal solution.

References

1. Fogel, L.J., Owens, A.J., Walsh, M.J.: Artificial Intelligence Through Simulated Evolution. Wiley, New York (1966)
2. Helbig, M., Engelbrecht, A.: Dynamic multi-objective optimization using pso. Metaheuristics Dyn. Optim. **433**, 147–188 (2013)
3. Trojanowski, K., Wierzchon, S.: Immune-based algorithms for dynamic optimization. Inf. Sci. **179**(10), 1495–1515 (2009)
4. Liu, R., Fan, J., Jiao, L.: Integration of improved predictive model and adaptive differential evolution based dynamic multi-objective evolutionary optimization algorithm. Appl. Intell. **43**(1), 192–207 (2015)
5. Jin, Y., Branke, J.: Evolutionary optimization in uncertain environments - a survey. IEEE Trans. Evol. Comput. **9**(3), 303–317 (2005)
6. Deb, K., Rao, U., Karthik, S.: Dynamic multi-objective optimization and decision-making using modified nsga-ii: a case study on hydro-thermal power scheduling. In: Obayashi, S., et al. (eds.) Proceedings of the 4th International Conference, EMO 2007, vol. 4403, pp. 803–817 (2007)
7. Azzouz, R., Bechikh, S., Said, L.B.: Multi-objective optimization with dynamic constraints and objectives: new challenges for evolutionary algorithms. In: Genetic and Evolutionary Computation Conference (GECCO 2015), pp. 615–622 (2015)
8. Hatzakis, I., Wallace, D.: Dynamic multi-objective optimization with evolutionary algorithms: a forward-looking approach. In: Proceedings of the 2006 Genetic and Evolutionary Computation Conference, pp. 1201–1208 (2006)
9. Koo, W.T., Goh, C., Tan, K.: A predictive gradient strategy for multi-objective evolutionary algorithms in a fast changing environment. Memet. Comput. **2**(2), 87–110 (2010)

10. Goh, C.K., Tan, K.C.: A competitive-cooperative coevolutionary paradigm for dynamic multi-objective optimization. IEEE Trans. Evol. Comput. **13**(1), 103–127 (2009)
11. Cámara, M., Ortega, J., de Toro, F.: Parallel processing for multi-objective optimization in dynamic environments. In: Proceedings of the IEEE International Parallel and Distributed Processing Symposium, pp. 1–8 (2007)
12. Shengxiang, Y., Soon Ong, Y., Jin, Y.: Evolutionary Computation in Dynamic and Uncertain Environments. Studies in Computational Intelligence, vol. 51. Springer, Berlin (2007)
13. Cruz, C., Gonzalez, J.R., Pelta, D.A.: Optimization in dynamic environments: a survey on problems, methods and measures. Soft Comput. **15**(7), 1427–1448 (2011)
14. Helbig, M., Engelbrecht, A.P.: Population-based metaheuristics for continuous boundary-constrained dynamic multi-objective optimisation problems. Swarm Evol. Comput. **14**, 31–47 (2014)
15. Carlo, R., Xin, Y.: Dynamic Multi-objective Optimization: A survey of the state-of-the-Art. Evolutionary Computation for Dynamic and Optimization Problems, pp. 85–106. Springer, Berlin (2013)
16. Hendrik, R.: Dynamic fitness landscape analysis. Evol. Comput. Dyn. Optim. Probl. **490**, 269–297 (2013)
17. Farina, M., Amato, P., Deb, K.: Dynamic multi-objective optimization problems: test cases, approximations and applications. IEEE Trans. Evol. Comput. **8**(5), 425–442 (2004)
18. Grefenstette, J.J.: Genetic algorithms for changing environments. In: Proceedings of the Second International Conference on Parallel Problem Solving from Nature, pp. 137–144 (1992)
19. Yang, S.: Genetic algorithms with memory and elitism-based immigrants in dynamic environment. Evol. Comput. **16**(3), 385–416 (2008)
20. Cobb, H.G.: An investigation into the use of hypermutation as an adaptive operator in genetic algorithms having continuous, time-dependent nonstationary environments. Technical Report AIC-90-001, Naval Research Laboratory (1990)
21. Morrison, R.W., Jon, K.A.D.: Triggered hypermutation revisited. Proc. IEEE Congr. Evol. Comput. **2**, 1025–1032 (2000)
22. Ramsey, C.L., Grefenstette, J.J.: Case-based initialization of genetic algorithms. In: Proceedings of the 5th International Conference on Genetic Algorithms, pp. 84–91 (1993)
23. Yang, S., Yao, X.: Population-based incremental learning with associative memory for dynamic environments. IEEE Trans. Evol. Comput. **12**(5), 542–561 (2008)
24. Oppacher, F., Wineberg, M.: The shifting balance genetic algorithm: improving the ga in a dynamic environment. Proc. Genet. Evol. Comput. Conf. **1**, 504–510 (1999)
25. Li, C., Yang, S.: A general framework of multipopulation methods with clustering in undetectable dynamic environments. IEEE Trans. Evol. Comput. **16**(4), 556–577 (2012)
26. Bosman, P.A.N.: Learning and anticipation in online dynamic optimization. In: Evolutionary Computation in Dynamic and Uncertain Environments, pp. 129–152 (2007)
27. Zhang, Q.F., Zhou, A.M., Jin, Y.C.: Rm-meda: a regularity model-based multi-objective estimation of distribution algorithm. IEEE Trans. Evol. Comput. **12**(1), 41–63 (2008)
28. Deb, K.: Multi-objective genetic algorithms: problem difficulties and construction of test problems. Evol. Comput. **7**(3), 205–230 (1999)
29. Woldesenbet, Y.G., Yen, G.G., Tessema, B.: Constraint handling in multi-objective evolutionary optimization. IEEE Trans. Evol. Comput. **13**(3), 514–525 (2009)
30. Chen, H., Li, M., Chen, X.: Using diversity as an additional-objective in dynamic multi-objective optimization algorithms. In: Second International Symposium on Electronic Commerce and Security, ISECS '09, vol. 1, pp. 484–487 (2009)
31. van Veldhuizen, D.A.: Multi-objective evolutionary algorithms: classification, analyses, and new innovations. Ph.D. thesis, Graduate School of engineering Air University (1999)
32. Sierra M., Coello, C.C.: Improving pso-based multi-objective optimization using crowding, mutation and epsilon-dominance. In: 3rd International Conference On Evolutionary multi-criterion optimization, vol. 3410, pp. 505–519 (2005)
33. Mehnen, J., Wagner, T., Rudolph, G.: Evolutionary optimization of dynamic multi-objective test functions. In: Proceedings of the second Italian Workshop on Evolutionary Computation (2006)

34. Zhou, A., Jin, Y.C., Zhang, Q., Sendhoff, B., Tsang, E.: Prediction-based population re-initialization for evolutionary dynamic multi-objective optimization. In: Proceedings of the 4th International Conference on Evolutionary Multi-Criterion Optimization, pp. 832–846 (2007)
35. Li, H., Zhang, Q.: A multiobjective differential evolution based on decomposition for multiobjective optimization with variable linkages. Parallel Probl. Solving Nat. **4193**, 583–592 (2006)
36. Roy, R., Mehnen, J.: Dynamic multi-objective optimisation for machining gradient materials. CIRP Ann. Manuf. Technol. **57**(1), 429–432 (2008)
37. Liu, C.: New dynamic multiobjective evolutionary algorithm with core estimation of distribution. In: International Conference on Electrical and Control Engineering (ICECE), pp. 1345–1348 (2010)
38. Jin, Y., Sendhoff, B.: Constructing dynamic optimization test problems using the multi-objective optimization concept. In: Proceedings of the EvoWorkshops, pp. 525–536 (2004)
39. Helbig, M., Engelbrecht, A.P.: Benchmarks for dynamic multi-objective optimisation algorithms. ACM Comput. Surv. **46**(3), 37:1–37:39 (2014)
40. Zhou, A., Jin, Y., Zhang, Q.: A population prediction strategy for evolutionary dynamic multiobjective optimization. IEEE Trans. Cybern. **44**(1), 40–53 (2014)
41. Li, Z., Chen, H., Xie, Z., Chen, C., Sallam, A.: Dynamic multiobjective optimization algorithm based on average distance linear prediction model. Sci. World J. **2014**, 9 (2014)
42. Muruganantham, A., Tan, K.C., Vadakkepat, P.: Solving the ieee cec 2015 dynamic benchmark problems using kalman filter based dynamic multiobjective evolutionary algorithm. Intell. Evol. Syst. **5**, 239–252 (2015)
43. Hatzakis, I., Wallace, D.: Topology of anticipatory populations for evolutionary dynamic multi-objective optimization. In: Proceedings of the 11th AIAA/ISSMO Multidisciplinary Analysis and Optimization Conference (2006)
44. Tan, K., Chew, Y., Lee, T., Yang, Y.: A cooperative coevolutionary algorithm for multiobjective optimization. IEEE Int. Conf. Syst. Man Cybern. **1**, 390–395 (2003)
45. Knowles, J., Corne, D.: The pareto archived evolution strategy: a new baseline algorithm for pareto multiobjective optimisation. In: Proceedings of the 1999 Congress on Evolutionary Computation, CEC 99, vol. 1, p. 105, (1999)
46. Leung, Y.-W., Wang, Y.: U-measure: a quality measure for multiobjective programming. IEEE Trans. Syst. Man Cybern. Part A: Syst. Hum. **33**(3), 337–343 (2003)
47. Wang, Y., Li, B.: Multi-strategy ensemble evolutionary algorithm for dynamic multi-objective optimization. Memet. Comput. **2**(1), 3–24 (2010)
48. Wang, Y., Li, B.: Investigation of memory-based multi-objective optimization evolutionary algorithm in dynamic environment. In: Proceedings of the IEEE Congress on Evolutionary Computation, pp. 630–637 (2009)
49. Deb, K., Pratap, A., Agarwal, S., Meyarivan, T.: A fast and elitist multiobjective genetic algorithm: Nsga-ii. IEEE Trans. Evol. Comput. **6**(2), 182–197 (2002)
50. Azzouz, R., Bechikh, S., Said, L.B.: A dynamic multi-objective evolutionary algorithm using a change severity-based adaptive population management strategy. In: Soft Computing, pp. 1–22 (2015)
51. Zitzler, E., Thiele, L.: Multiobjective evolutionary algorithms: a comparative case study and the strength pareto approach. IEEE Trans. Evol. Comput. **3**(4), 257–271 (1999)
52. Alba, E.: Parallel evolutionary algorithms can achieve super-linear performance. Inf. Process. Lett. **82**(1), 7–13 (2002)
53. Zheng, B.: A new dynamic multi-objective optimization evolutionary algorithm. In: Proceedings of the Third International Conference on Natural Computation, pp. 565–570 (2007)
54. Cámara, M., Ortega, J., de Toro, F.: Parallel multi-objective optimization evolutionary algorithms in dynamic environments. Proc. First Int. Workshop Parallel Archit. Bioinspired Algorithms **1**, 13–20 (2008)
55. Wang, Y., Dang, C.: An evolutionary algorithm for dynamic multi-objective optimization. Appl. Math. Comput. **205**(1), 6–18 (2008)

56. Liu, C.-A., Wang, Y.: New evolutionary algorithm for dynamic multiobjective optimization problems. Adv. Nat. Comput. **4221**, 889–892 (2006)
57. Liu, C.-A., Wang, Y.: Dynamic multi-objective optimization evolutionary algorithm. In: Third International Conference on Natural Computation, ICNC 2007, vol. 4, pp. 456–459 (2007)
58. Liu, C.A., Wang, Y., Ren, A.: New dynamic multi-objective constrained optimization evolutionary algorithm. Asia-Pac. J. Oper. Res. **32**(05) (2015)
59. Zitzler, E.: Evolutionary algorithms for multiobjective optimization: methods and applications. Ph.D. thesis, Swiss Federal Institute of Technology (ETH) Zurich, Switzerland (1999)
60. Guan, S.U., Chen, Q., Mo, W.: Evolving dynamic multi-objective optimization problems with objective replacement. Artif. Intell. Rev. **23**(3), 267–293 (2005)
61. Zeng, S., Yao, S., Kang, L., Liu, Y.: An efficient multi-objective evolutionary algorithm: Omoea-ii. In: Third International Conference on Evolutionary Multi-criterion Optimization (EMO 2005), pp. 108–119 (2005)
62. Amato, P., Farina, M.: An alife-inspired evolutionary algorithm for dynamic multi-objective optimization problems. Adv. Soft Comput. **32**, 113–125 (2005)
63. Zeng, S.Y., Chen, G., Zheng, L., Shi, H., de Garis, H., Ding, L., Kang, L.: A dynamic multi-objective evolutionary algorithm based on an orthogonal design. In: Proceedings of the IEEE Congress on Evolutionary Computation, pp. 573–580 (2006)
64. Deb, K.: Single and multi-objective dynamic optimization: two tales from an evolutionary perspective. Technical Report 2011004, Kanpur Genetic Algorithms Laboratory (2011)
65. Huang, L., Suh, I., Abraham, A.: Dynamic multi-objective optimization based on membrane computing for control of time-varying unstable plants. Inf. Sci. **181**(11), 2370–2391 (2011)
66. Azzouz, R., Bechikh, S., Said, L.B.: A multiple reference point-based evolutionary algorithm for dynamic multi-objective optimization with undetectable changes. In: Proceedings of the IEEE Congress on Evolutionary Computation, pp. 3168–3175 (2014)
67. Xiaodong, L., Branke, J., Kirley, M.: On performance metrics and particle swarm methods for dynamic multiobjective optimization problems. IEEE Congr. Evol. Comput. CEC **2007**, 576–583 (2007)
68. Liu, C.-A.: New dynamic multiobjective evolutionary algorithm with core estimation of distribution. In: International Conference on Electrical and Control Engineering (ICECE), pp. 1345–1348 (2010)
69. Tang, G.C.M., Huang, Z.: The construction of dynamic multi-objective optimization test functions. Adv. Comput. Intell. **4683**, 72–79 (2007)
70. Avdagic, S.O.Z., Konjicija, S.: Evolutionary approach to solving non-stationary dynamic multi-objective problems. Found. Comput. Intell. **3**(203), 267–289 (2009)
71. Helbig, M., Engelbrecht, A.: Archive management for dynamic multi-objective optimisation problems using vector evaluated particle swarm optimisation. In: IEEE Congress on Evolutionary Computation (CEC), pp. 2047–2054 (2011)
72. Helbig, M., Engelbrecht, A.: Benchmarks for dynamic multi-objective optimisation. In: IEEE Symposium on Computational Intelligence in Dynamic and Uncertain Environments (CIDUE), pp. 84–91 (2013)
73. Biswas, S., Das, S., Suganthan, P., Coello, C.C.: Evolutionary multiobjective optimization in dynamic environments: a set of novel benchmark functions. In: 2014 IEEE Congress on Evolutionary Computation (CEC), pp. 3192–3199 (2014)
74. Hamalainen, R.P., Mantysaari, J.: A dynamic interval goal programming approach to the regulation of a lake - river system. J. Multi-criteria Decis. Anal. **10**, 75–86 (2001)
75. Hamalainen, R.P., Mantysaari, J.: Dynamic multi-objective heating optimization. Eur. J. Oper. Res. **142**(1), 1–15 (2002)
76. Ursem, R., Krink, T., Filipic, B.: A numerical simulator for a crop-producing greenhouse. In: EVALife Technical Report, no. 2002-01 (2002)
77. Shen, X.-N., Yao, X.: Mathematical modeling and multi-objective evolutionary algorithms applied to dynamic flexible job shop scheduling problems. Inf. Sci. **298**, 198–224 (2015)
78. Nguyen, S., Zhang, M., Tan, K.C.: Enhancing genetic programming based hyper-heuristics for dynamic multi-objective job shop scheduling problems. In: 2015 IEEE Congress on Evolutionary Computation (CEC), pp. 2781–2788 (2015)

79. Palaniappan, S., Zein-Sabatto, S., Sekmen, A.: Dynamic multiobjective optimization of war resource allocation using adaptive genetic algorithms. In: Proceedings of IEEE SoutheastCon, pp. 160–165 (2001)
80. Hutzschenreuter, A., Bosman, P., Poutré, H.: Evolutionary multiobjective optimization for dynamic hospital resource management. In: Proceedings of International Conference on Multi-criterion Optimization, pp. 320–334 (2009)
81. Wahle, J., Annen, O., Schuster, C., Neubert, L., Schreckenberg, M.: A dynamic route guidance system based on real traffic data. Eur. J. Oper. Res. **131**(2), 302–308 (2001)
82. Constantinou, D.: Ant colony optimisation algorithms for solving multi-objective power aware metrics for mobile ad hoc networks. Ph.D. thesis, Department of Computer Science, University of Pretoria, South Africa (2011)
83. Grimme, C., Meisel, S., Trautmann, H., Rudolph, G., Wölck, M.: Multi-objective analysis of approaches to dynamic routing of a vehicle. In: Twenty-Third European Conference on Information Systems Completed Research Papers. Paper 62 (2015)
84. Meisel, S., Grimme, C., Bossek, J., Wölck, M., Rudolph, G., Trautmann, H.: Evaluation of a multi-objective ea on benchmark instances for dynamic routing of a vehicle. In: Proceedings of the 2015 Annual Conference on Genetic and Evolutionary Computation, pp. 425–432 (2015)
85. Chen, C.-L., Lee, W.-C.: Multi-objective optimization of multi-echelon supply chain networks with uncertain product demands and prices. Comput. Chem. Eng. **28**, 1131–1144 (2004)
86. Selim, H., Araz, C., Ozkarahan, I.: Collaborative production-distribution planning in supply chain: a fuzzy goal programming approach. Transp. Res. Part E: Logist. Transp. Rev. **44**(3), 396–419 (2008)
87. Maalawi, K.: Special issue on design optimization of wind turbine structures. In: *Wind Turbines* (2011)
88. Zhang, Z., Qian, S.: Artificial immune system in dynamic environments solving time-varying non-linear constrained multi-objective problems. Soft Comput. **15**(7), 1333–1349 (2011)
89. Weicker, K.: Performance measures for dynamic environments. In: Parallel Problem Solving from Nature, pp. 64–73 (2002)
90. Cámara, M., Ortega, J., Toro, F.d.: Approaching dynamic multi-objective optimization problems by using parallel evolutionary algorithms. In: Advances in Multi-objective Nature Inspired Computing, vol. 272, pp. 63–86 (2010)
91. Bechikh, S., Kessentini, M., Said, L.B., Ghedira, K.: Preference incorporation in evolutionary multiobjective optimization: a survey of the state-of-the-art. Advances in Computers, vol. 98, pp. 141–207. Elsevier (2015)
92. Bechikh, S.: Incorporating Decision Maker's Preference Information in Evolutionary Multi-objective Optimization. Ph.D. thesis, University of Tunis, ISG-Tunis, Tunisia (2013)
93. Bechikh, S., Said, L.B., Ghedira, K.: Negotiating decision makers' reference points for group preference-based evolutionary multi-objective optimization. In: 2011 11th International Conference on Hybrid Intelligent Systems (HIS), pp. 377–382 (2011)
94. Bechikh, S., Said, L.B., Ghedira, K.: Group preference-based evolutionary multi-objective optimization with non-equally important decision makers: Application to the portfolio selection problem. Int. J. Comput. Inf. Syst. Ind. Manag. Appl. **5**(1), 278–288 (2013)
95. Trabelsi, W., Azzouz, R., Bechikh, S., Said, L.B.: Leveraging evolutionary algorithms for dynamic multi-objective optimization scheduling of multi-tenant smart home appliances. In: IEEE Congress on Evolutionary Computation (2016)
96. Azzouz, R., Bechikh, S., Said, L.B.: Articulating decision maker's preference information within multiobjective artificial immune systems. In: 2012 IEEE 24th International Conference on Tools with Artificial Intelligence, vol. 1, pp. 327–334 (2012)
97. Bechikh, S., Chaabani, A., Said, L.B.: An efficient chemical reaction optimization algorithm for multiobjective optimization. IEEE Trans. Cybern. **45**(10), 2051–2064 (2015)

Evolutionary Bilevel Optimization: An Introduction and Recent Advances

Ankur Sinha, Pekka Malo and Kalyanmoy Deb

Abstract Bilevel optimization involves two levels of optimization where one optimization level acts as a constraint to another optimization level. There are enormous applications that are bilevel in nature; however, given the difficulties associated with solving this difficult class of problem, the area still lacks efficient solution methods capable of handling complex application problems. Most of the available solution methods can either be applied to highly restrictive class of problems, or are computationally very expensive such that they do not scale for large scale bilevel problems. The difficulties in bilevel programming arise primarily from the nested structure of the problem. Evolutionary algorithms have been able to demonstrate its potential in solving single-level optimization problems. In this chapter, we provide an introduction to the progress made by the evolutionary computation community towards handling bilevel problems. The chapter highlights past research and future research directions both on single as well as multiobjective bilevel programming. Some of the immediate application areas of bilevel programming have also been highlighted.

Keywords Bilevel optimization · Stackelberg games · Evolutionary algorithms · Mathematical programming

---

A. Sinha (✉)
Production and Quantitative Methods, Indian Institute of Management, Ahmedabad 380015, India
e-mail: asinha@iima.ac.in

P. Malo
Department of Information and Service Economy, Aalto University School of Business, PO Box 21220, 00076 Aalto, Finland
e-mail: pekka.malo@aalto.fi

K. Deb
Department of Electrical and Computer Engineering, Michigan State University, East Lansing, MI, USA
e-mail: kdeb@egr.msu.edu

S. Bechikh et al. (eds.), *Recent Advances in Evolutionary Multi-objective Optimization*, Adaptation, Learning, and Optimization 20,
DOI 10.1007/978-3-319-42978-6_3

1 Introduction

Bilevel optimization is characterized as a mathematical program with two levels of optimization. The outer optimization problem is commonly referred to as the upper level optimization problem and the inner optimization problem is commonly referred to as the lower level optimization problem. The origin of bilevel optimization can be traced to two roots: these problems were first realized by Stackelberg [1] in the area of game theory and came to be known as Stackelberg games; later these problems were realized in the area of mathematical programming by Bracken and McGill [2] as a constrained optimization task, where the lower level optimization problem acts as a constraint to the upper level optimization problem. These problems are known to be difficult due to its nested structure; therefore, it has received most attention from the mathematical community towards simple cases where the objective functions and constraints are linear [3, 4], quadratic [5–7] or convex [8]. The nested structure in bilevel introduces difficulties such as non-convexity and disconnectedness even for simpler instances of bilevel optimization like bilevel linear programming problems. Bilevel linear programming is known to be strongly NP-hard [9], and it has been proven that merely evaluating a solution for optimality is also a NP-hard task [10]. This gives us an idea about the kind of challenges offered by bilevel problems with complex (non-linear, non-convex, discontinuous etc.) objective and constraint functions.

An interest in bilevel programming has been driven by a number of new applications arising in different fields of optimization. For instance, in the context of homeland security [11–13], bilevel and even trilevel optimization models are common. In game theoretic settings, bilevel programs have been used in the context of optimal tax policies [14–16]; model production processes [17]; investigation of strategic behavior in deregulated markets [18] and optimization of retail channel structures [19], among others. Bilevel optimization applications are ubiquitous and arise in many other disciplines, like in transportation [20–22], management [23, 24], facility location [23, 25, 26], chemical engineering [27, 28], structural optimization [29, 30], and optimal control [31, 32] problems.

Evolutionary computation [33] techniques have been successfully applied to handle mathematical programming problems and applications that do not adhere to regularities like continuity, differentiability or convexities. Due to these properties of evolutionary algorithms, attempts have been made to solve bilevel optimization problems using these methods, as even simple (linear or quadratic) bilevel optimization problems are intrinsically non-convex, non-differentiable and disconnected at times. However, the advantages come with a trade-off. Most of the evolutionary bilevel techniques are nested where an outer algorithm handles the upper level optimization task and an inner algorithm handles the lower level optimization task, thereby making the overall bilevel optimization computationally very intensive. To address these problems attempts have been made to reduce the computational expense of evolutionary bilevel optimization algorithms by utilizing metamodeling-based principles. Multiobjective bilevel programming is a natural extension of bilevel optimization problems with single objectives. However, multiple objectives in bilevel optimization, along with computational challenges, brings in intricacies related to hierarchical decision making.

In this chapter, we highlight some of the past, and recent studies and results in the area of evolutionary bilevel optimization. The chapter begins with a survey on single objective bilevel optimization in Sect. 2. This is followed by single-level formulations of bilevel optimization in Sect. 3. Thereafter, in Sect. 4 we discuss and compare some recent solution methods for bilevel optimization. Section 5 introduces multiobjective bilevel optimization and provides a survey on the topic. In Sect. 6 we discuss the decision making issues in multiobjective bilevel optimization. Finally, we conclude in Sect. 7 with some ideas on future research directions.

2 A Survey on Evolutionary Bilevel Optimization

Most of the evolutionary approaches proposed to handle bilevel optimization problems are nested in nature. As the name suggests, these approaches rely on two optimization algorithms, where one algorithm is executed within the other. Based on the complexity of the optimization tasks at each level, researchers have chosen to use either evolutionary algorithms at both levels or evolutionary algorithm at one level and classical optimization algorithm at the other level. One of the earliest evolutionary algorithms for solving bilevel optimization problems was proposed in the early 1990s by Mathieu et al. [34] who used a nested approach with genetic algorithm at the upper level, and linear programming at the lower level. Later, Yin [35] used genetic algorithm at the upper level and Frank–Wolfe algorithm (reduced gradient method) at the lower level. In both these approaches a lower level optimization task was executed for every upper level member that emphasizes the nested structure of these approaches. Along similar lines, nested procedures were used in [36–39]. Approaches with evolutionary algorithms at both levels are also common; for instance, in [40] authors used differential evolution at both levels, and in [41] authors nested differential evolution within an ant colony optimization.

In a number of studies, where lower level problem adhered to certain regularity conditions, researchers have used the KKT conditions for the lower level problem to reduce the bilevel problem into a single-level problem. The reduced single-level problem is then solved with an evolutionary algorithm. For instance, Hejazi et al. [42], reduced the linear bilevel problem to single-level and then used a genetic algorithm, where chromosomes emulate the vertex points, to solve the problem. Wang et al. [43] used KKT conditions to reduce the bilevel problem into single-level, and then utilized a constraint handling scheme to successfully solve a number of standard test problems. A later study by Wang et al. [44] introduced an improved algorithm that performed better than the previous approach [43]. Recently, Jiang et al. [45] reduced the bilevel optimization problem into a non-linear optimization problem with complementarity constraints, which is sequentially smoothed and solved with a PSO algorithm. Other studies using similar ideas are [46, 47].

It is noteworthy that utilization of KKT conditions restricts the algorithm's applicability to only a special class of bilevel problems. To overcome this drawback, researchers are looking into metamodeling based approaches where the lower level optimal reaction set is approximated over generations of the evolutionary algorithm. Studies in this direction are [48, 49]. Along similar lines, attempts have been made to metamodel the lower level optimal value function [50] to solve bilevel optimization problems. Approximating the lower level optimal value function may offer a few advantages over approximating the lower level reaction set that has been highlighted in this chapter.

3 Bilevel Formulation and Single-Level Reductions

In this section, we provide a general formulation for bilevel optimization, and different ways people have used to reduce bilevel optimization problems to single-level problems. Bilevel problems contain two levels, upper and lower, where lower level is nested within the upper level problem. The two levels have their own objectives, constraints and variables. In the context of game theory, the two problems are also referred to as the leader's (upper) and follower's problems (lower). The lower level optimization problem is a parametric optimization problem that is solved with respect to the lower level variables while the upper level variables act as parameters. The difficulty in bilevel optimization arises from the fact that only lower level optimal solutions can be considered as feasible members, if they also satisfy the upper level constraints. Below we provide a general bilevel formulation:

Definition 1 For the upper-level objective function $F : \mathbb{R}^n \times \mathbb{R}^m \to \mathbb{R}$ and lower-level objective function $f : \mathbb{R}^n \times \mathbb{R}^m \to \mathbb{R}$, the bilevel optimization problem is given by

$$\begin{aligned} \text{“min”}_{x_u \in X_U, x_l \in X_L} \quad & F(x_u, x_l) \text{ subject to} \\ & x_l \in \underset{x_l \in X_L}{\operatorname{argmin}}\{f(x_u, x_l) : g_j(x_u, x_l) \le 0, j = 1, \ldots, J\} \\ & G_k(x_u, x_l) \le 0, k = 1, \ldots, K \end{aligned}$$

where $G_k : X_U \times X_L \to \mathbb{R}$, $k = 1, \ldots, K$ denotes the upper level constraints, and $g_j : X_U \times X_L \to \mathbb{R}$ represents the lower level constraints, respectively.

3.1 *Optimistic Versus Pessimistic*

Quotes have been used while specifying the upper level minimization problem in Definition 1 because the problem is ill-posed for cases where the lower level has multiple optimal solutions. For instance, Fig. 1 shows the case where the lower level

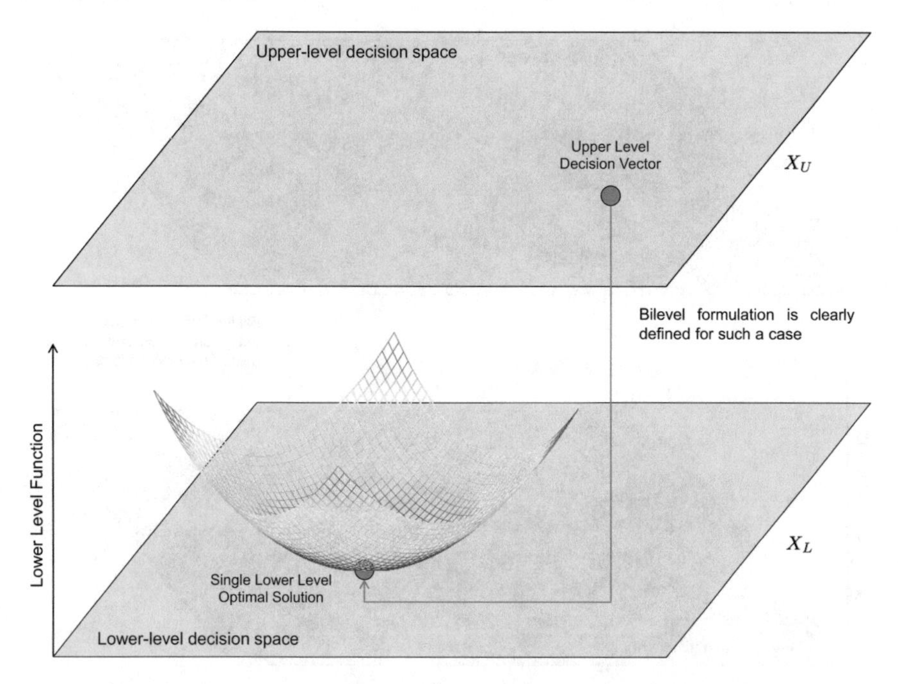

Fig. 1 A scenario where there is a single lower level optimal solution corresponding to an upper level decision vector. The bilevel optimization problem in Definition 1 is clearly defined for this case

problem has a single optimal solution corresponding to an upper level decision. Therefore, it is clear that for the upper level decision, the only rational lower level decision would be the single optimal solution at the lower level. However, there is lack of clarity in the situation shown in Fig. 2, as it is not clear that, out of multiple lower level optimal solutions, which solution will actually be chosen by the lower level decision maker. If the selection of the lower level decision maker is unknown, the bilevel formulation remains ill-defined. It is common to assume either of the two positions, i.e., optimistic or pessimistic, to sort out this ambiguity. In optimistic position some form of cooperation is assumed between the leader and the follower. For any given leader's decision vector that has multiple optimal solutions for the follower, the follower is expected to choose that optimal solution that leads to the best objective function value for the leader. On the other hand, in a pessimistic position the leader optimizes for the worst case, i.e. the follower may choose that solution from the optimal set which leads to the worst objective function value for the leader. Optimistic position being more tractable is commonly studied in the literature, and we also consider the optimistic position in this chapter.

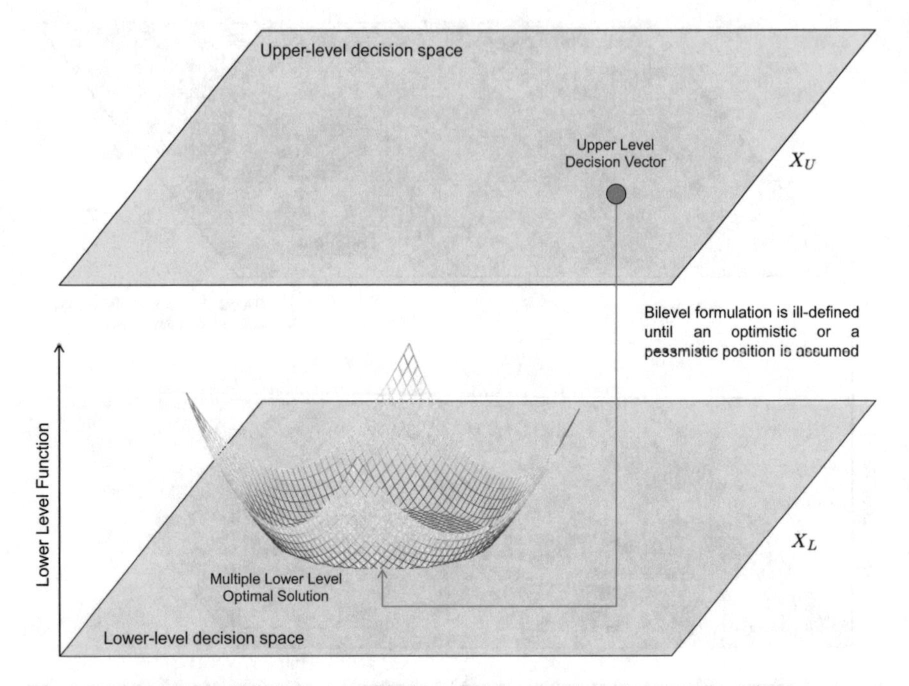

Fig. 2 A scenario where there is a multiple lower level optimal solution corresponding to an upper level decision vector. The bilevel optimization problem in Definition 1 is ill-defined for this case if the lower level's selection is not known or assumed

3.2 KKT Reduction

When the lower level problem in Definition 1 adheres to certain convexity and regularity conditions, it is possible to replace the lower level optimization task with its KKT conditions.

Definition 2 The KKT conditions appear as Lagrangian and complementarity constraints in the single-level formulation provide below:

$$
\begin{aligned}
&\min_{x_u \in X_U, x_l \in X_L, \lambda} \quad F(x_u, x_l) \\
&\text{subject to} \\
&\qquad G_k(x_u, x_l) \leq 0, \; k = 1, \ldots, K, \\
&\qquad g_j(x_u, x_l) \leq 0, \; j = 1, \ldots, J, \\
&\qquad \lambda_j g_j(x_u, x_l) = 0, \; j = 1, \ldots, J, \\
&\qquad \lambda_j \geq 0, \; j = 1, \ldots, J, \\
&\qquad \nabla_{x_l} L(x_u, x_l, \lambda) = 0,
\end{aligned}
$$

where

$$L(x_u, x_l, \lambda) = f(x_u, x_l) + \sum_{j=1}^{J} \lambda_j g_j(x_u, x_l).$$

The above formulation might not be simple to handle, as the Lagrangian constraints often lead to non-convexities, and the complementarity condition being combinatorial, make the overall problem a mixed integer problem. In case of linear bilevel optimization problems, the Lagrangian constraint is also linear. Therefore, the single-level reduced problem becomes a mixed integer linear program. Approaches based on vertex enumeration [51–53], as well as branch-and-bound [54, 55] have been proposed to solve these problems.

3.3 Reaction Set Mapping

An equivalent formulation of the problem given in Definition 1 can be stated in terms of set-valued mappings as follows:

Definition 3 Let $\Psi : \mathbb{R}^n \rightrightarrows \mathbb{R}^m$ be the reaction set mapping,

$$\Psi(x_u) = \underset{x_l \in X_L}{\operatorname{argmin}}\{f(x_u, x_l) : g_j(x_u, x_l) \leq 0, j = 1, \ldots, J\},$$

which represents the constraint defined by the lower-level optimization problem, i.e. $\Psi(x_u) \subset X_L$ for every $x_u \in X_U$. Then the bilevel optimization problem can be expressed as a constrained optimization problem as follows:

$$\begin{aligned} &\min_{x_u \in X_U, x_l \in X_L} \quad F(x_u, x_l) \\ &\text{subject to} \\ &\qquad x_l \in \Psi(x_u) \\ &\qquad G_k(x_u, x_l) \leq 0, k = 1, \ldots, K \end{aligned}$$

Note that if the Ψ-mapping can somehow be determined, the problem reduces to a single level constrained optimization task. However, that is rarely the case. Evolutionary computation studies that rely on iteratively mapping this set to avoid frequent lower level optimization are [48, 49]. The idea behind the algorithm has been shown through Figs. 3 and 4. To begin with, the lower level problem is completely solved for a few upper level decision vectors. For example, in Fig. 3 the lower level decisions corresponding to upper level decisions a, b, c, d, e and f are determined by solving a lower level problem completely. The lower level decisions for these members correspond to the actual Ψ-mapping (unknown). These member are then used to find an approximate Ψ-mapping locally as shown in Fig. 4. For every new upper level

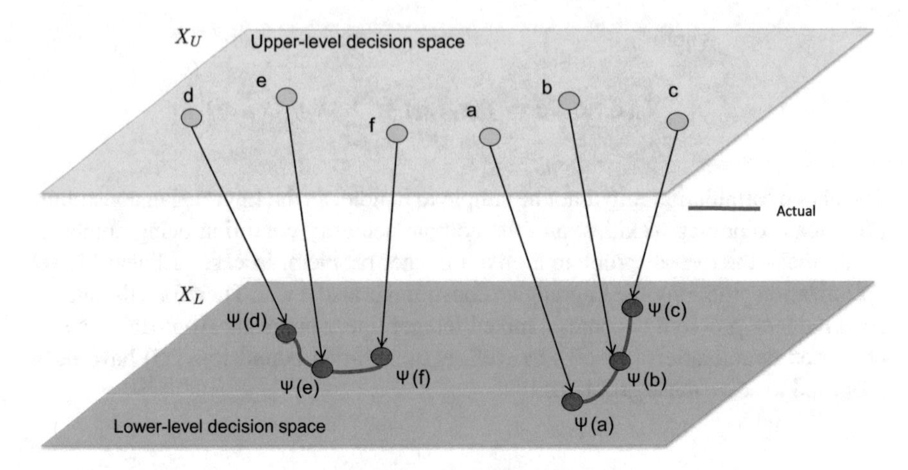

Fig. 3 Solving the lower level optimization problem completely for random upper level members like a, b, c, d, e and f provides the corresponding lower level optimal solutions represented by $\Psi(a)$, $\Psi(b)$, $\Psi(c)$, $\Psi(d)$, $\Psi(e)$ and $\Psi(f)$. The Ψ-mapping is assumed to be single valued

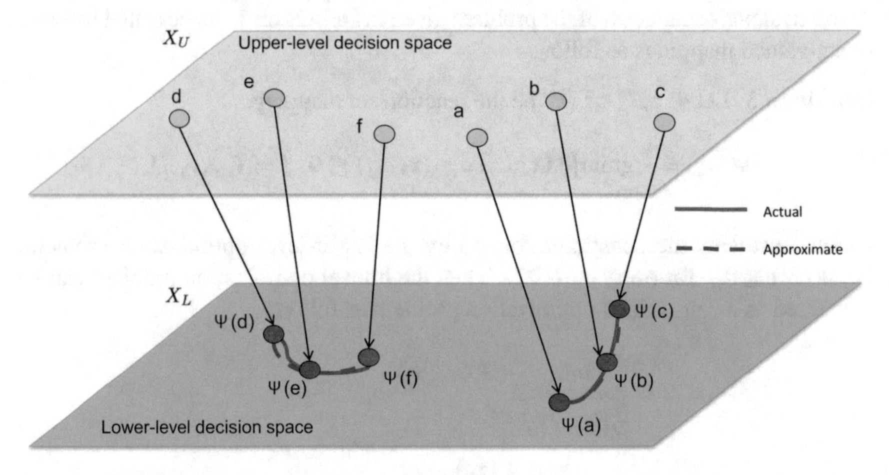

Fig. 4 An approximate mapping for the lower level reaction set estimated using the actual values $\Psi(a)$, $\Psi(b)$, $\Psi(c)$, $\Psi(d)$, $\Psi(e)$ and $\Psi(f)$. Local approximations are preferable over a global approximation of the Ψ-mapping

member, the local approximation is used to identify the lower level decision instead of solving the lower level optimization problem. The idea is used iteratively until convergence. The idea works well when the Ψ-mapping is single valued.

3.4 Lower Level Optimal Value Function

Another equivalent definition of the problem in Definition 1 can be given in terms of the lower level optimal value function that is defined below [56]:

Definition 4 Let $\varphi : X_U \rightarrow R$ be the lower level optimal value function mapping,

$$\varphi(x_u) = \min_{x_l \in X_L} \{f(x_u, x_l) : g_j(x_u, x_l) \leq 0, j = 1, \ldots, J\},$$

which represents the minimum lower level function value corresponding to any upper level decision vector. Then the bilevel optimization problem can be expressed as follows:

$$\begin{aligned} &\min_{x_u \in X_U, x_l \in X_L} \quad F(x_u, x_l) \\ &\text{subject to} \\ &\quad f(x_u, x_l) \leq \varphi(x_u) \\ &\quad g_j(x_u, x_l) \leq 0, j = 1, \ldots, J \\ &\quad G_k(x_u, x_l) \leq 0, k = 1, \ldots, K. \end{aligned}$$

The φ-mapping can be approximated iteratively during the generations of the evolutionary algorithm, and a reduced problem described in Definition 4 can be frequently solved to converge towards the bilevel optimum. An evolutionary algorithm relying on this idea can be found in [50]. Approximating the optimal value function mapping offers an advantage over approximating reaction set mapping, as the optimal value function mapping is not set valued. Moreover, it returns a scalar for any given upper level decision vector. Figure 5 shows an example where the lower level problem has multiple optimal solutions for some upper level decisions and single optimal solutions for others. In all situations, the φ-mapping remains single valued scalar. Though there are advantages associated with estimating the φ-mapping, it is also interesting to note in Definition 4 that the reduced single level problem has to be

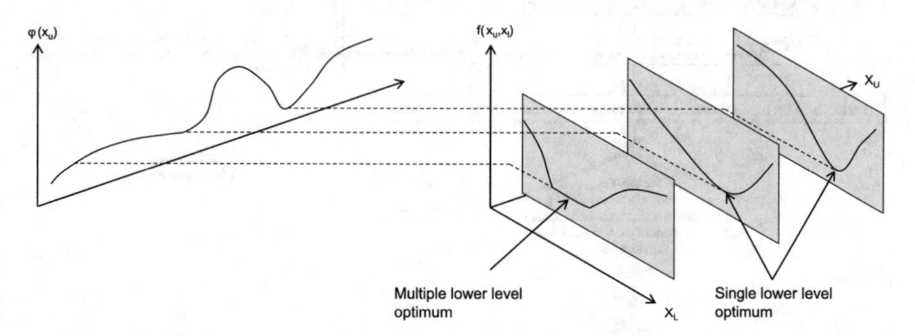

Fig. 5 An example showing φ-mapping and how it depends on the lower level optimization problem

solved with respect to both upper and lower level variables, while in Definition 7, the lower level variables are directly available from the Ψ-mapping. Therefore, there exists a trade-off.

4 Comparison of Metamodeling Based Evolutionary Approaches for Bilevel Optimization

In this section, we provide the steps of two different evolutionary bilevel algorithms, where one utilizes iterative approximation of the Ψ-mapping, while the other utilizes iterative approximation of the φ-mapping in the intermediate steps. The steps of the algorithms are provided through a flowchart in Fig. 6. For brevity, we do not discuss the steps of the evolutionary algorithm, as any scheme can be utilized in the provided framework to handle bilevel optimization problems. For further information about the implementation of the approaches the readers are referred to [50].

The intermediate steps of the above algorithms utilizes quadratic approximation for approximating the Ψ and the φ mappings. Both the ideas were tested on a set of 8 test problems given in Tables 1 and 2. To assess the savings achieved by the two approximation approaches, we compare them against a nested approach where the approximation idea is not incorporated, but the same evolutionary algorithm

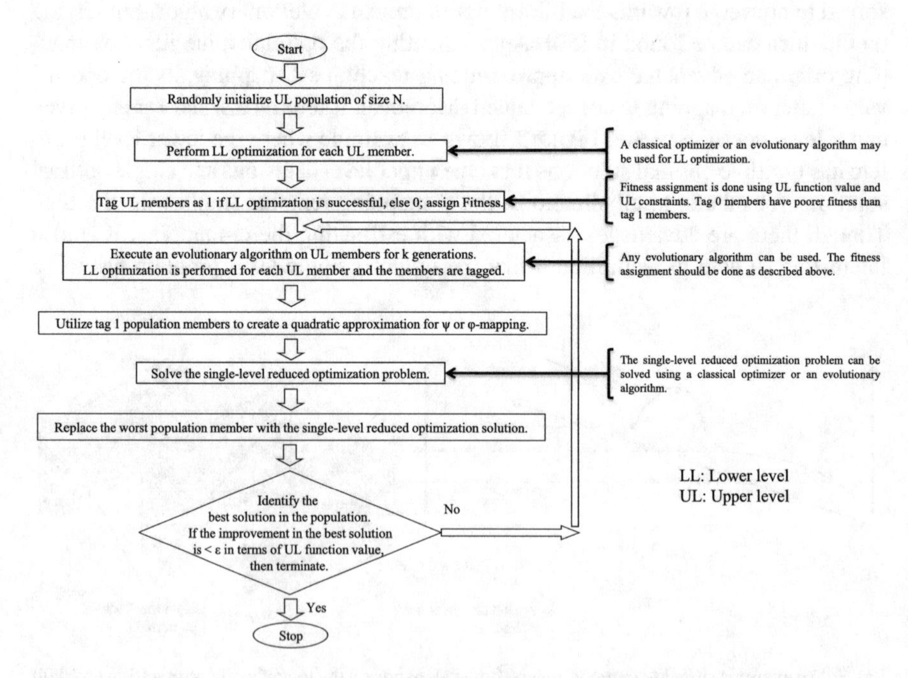

Fig. 6 Flowchart for incorporating approximated φ-mapping in an evolutionary algorithm

Table 1 Standard test problems TP1–TP5

Problem	Formulation	Best Known Sol.
TP1 $n=2$, $m=2$	$\underset{(x,y)}{\text{Minimize}}\ F(x,y)=(x_1-30)^2+(x_2-20)^2-20y_1+20y_2,$ s.t. $y\in\underset{(y)}{\text{argmin}}\left\{\begin{array}{l} f(x,y)=(x_1-y_1)^2+(x_2-y_2)^2\\ 0\le y_i\le 10,\quad i=1,2\end{array}\right\},$ $x_1+2x_2\ge 30,\ x_1+x_2\le 25,\ x_2\le 15$	$F=225.0$ $f=100.0$
TP2 $n=2$, $m=2$	$\underset{(x,y)}{\text{Minimize}}\ F(x,y)=2x_1+2x_2-3y_1-3y_2-60,$ s.t. $y\in\underset{(y)}{\text{argmin}}\left\{\begin{array}{l} f(x,y)=(y_1-x_1+20)^2+(y_2-x_2+20)^2\\ x_1-2y_1\ge 10,\ x_2-2y_2\ge 10\\ -10\ge y_i\ge 20,\quad i=1,2\end{array}\right\},$ $x_1+x_2+y_1-2y_2\le 40,$ $0\le x_i\le 50,\quad i=1,2.$	$F=0.0$ $f=100.0$
TP3 $n=2$, $m=2$	$\underset{(x,y)}{\text{Minimize}}\ F(x,y)=-(x_1)^2-3(x_2)^2-4y_1+(y_2)^2,$ s.t. $y\in\underset{(y)}{\text{argmin}}\left\{\begin{array}{l} f(x,y)=2(x_1)^2+(y_1)^2-5y_2\\ (x_1)^2-2x_1+(x_2)^2-2y_1+y_2\ge -3\\ x_2+3y_1-4y_2\ge 4\\ 0\le y_i,\quad i=1,2\end{array}\right\},$ $(x_1)^2+2x_2\le 4,$ $0\le x_i,\quad i=1,2$	$F=-18.6787$ $f=-1.0156$
TP4 $n=2$, $m=3$	$\underset{(x,y)}{\text{Minimize}}\ F(x,y)=-8x_1-4x_2+4y_1-40y_2-4y_3,$ s.t. $y\in\underset{(y)}{\text{argmin}}\left\{\begin{array}{l} f(x,y)=x_1+2x_2+y_1+y_2+2y_3\\ y_2+y_3-y_1\le 1\\ 2x_1-y_1+2y_2-0.5y_3\le 1\\ 2x_2+2y_1-y_2-0.5y_3\le 1\\ 0\le y_i,\quad i=1,2,3\end{array}\right\},$ $0\le x_i,\quad i=1,2$	$F=-29.2$ $f=3.2$

(Note that $x=x_u$ and $y=x_l$)

described in Fig. 6 is used at the upper level and a lower level optimization problem is solved for every upper level member. Hereafter, we refer this benchmark as a no-approximation approach. Whenever lower level optimization is required, we rely on sequential quadratic programming to solve the problem for all cases. Table 3 provides the median function evaluations (31 runs) at the upper and lower level required by each of the three cases, i.e., φ-approximation, Ψ-approximation and no-approximation. Detailed results from multiple runs are presented through Figs. 7 and 8. Interestingly, both the approximation ideas perform significantly well on all

Table 2 Standard test problems TP6–TP8

Problem	Formulation	Best Known Sol.
TP5 $n=2, m=2$	$\underset{(x,y)}{\text{Minimize}}\ F(x,y) = rt(x)x - 3y_1 - 4y_2 + 0.5t(y)y,$ s.t. $y \in \underset{(y)}{\text{argmin}} \left\{ \begin{array}{l} f(x,y) = 0.5t(y)hy - t(b(x))y \\ -0.333y_1 + y_2 - 2 \le 0 \\ y_1 - 0.333y_2 - 2 \le 0 \\ 0 \le y_i, \quad i = 1,2 \end{array} \right\},$ where $h = \begin{pmatrix} 1 & 3 \\ 3 & 10 \end{pmatrix}, b(x) = \begin{pmatrix} -1 & 2 \\ 3 & -3 \end{pmatrix} x, r = 0.1$ $t(\cdot)$ denotes transpose of a vector	$F = -3.6$ $f = -2.0$
TP6 $n=1, m=2$	$\underset{(x,y)}{\text{Minimize}}\ F(x,y) = (x_1-1)^2 + 2y_1 - 2x_1,$ s.t. $y \in \underset{(y)}{\text{argmin}} \left\{ \begin{array}{l} f(x,y) = (2y_1-4)^2 + \\ (2y_2-1)^2 + x_1y_1 \\ 4x_1 + 5y_1 + 4y_2 \le 12 \\ 4y_2 - 4x_1 - 5y_1 \le -4 \\ 4x_1 - 4y_1 + 5y_2 \le 4 \\ 4y_1 - 4x_1 + 5y_2 \le 4 \\ 0 \le y_i, \quad i = 1,2 \end{array} \right\},$ $0 \le x_1$	$F = -1.2091$ $f = 7.6145$
TP7 $n=2, m=2$	$\underset{(x,y)}{\text{Minimize}}\ F(x,y) = -\frac{(x_1+y_1)(x_2+y_2)}{1+x_1y_1+x_2y_2},$ s.t. $y \in \underset{(y)}{\text{argmin}} \left\{ \begin{array}{l} f(x,y) = \frac{(x_1+y_1)(x_2+y_2)}{1+x_1y_1+x_2y_2} \\ 0 \le y_i \le x_i, \quad i = 1,2 \end{array} \right\},$ $(x_1)^2 + (x_2)^2 \le 100$ $x_1 - x_2 \le 0$ $0 \le x_i, \quad i = 1,2$	$F = -1.96$ $f = 1.96$
TP8 $n=2, m=2$	$\underset{(x,y)}{\text{Minimize}}\ F(x,y) = \lvert 2x_1 + 2x_2 - 3y_1 - 3y_2 - 60 \rvert,$ s.t. $y \in \underset{(y)}{\text{argmin}} \left\{ \begin{array}{l} f(x,y) = (y_1 - x_1 + 20)^2 + \\ (y_2 - x_2 + 20)^2 \\ 2y_1 - x_1 + 10 \le 0 \\ 2y_2 - x_2 + 10 \le 0 \\ -10 \le y_i \le 20, \quad i = 1,2 \end{array} \right\},$ $x_1 + x_2 + y_1 - 2y_2 \le 40$ $0 \le x_i \le 50, \quad i = 1,2$	$F = 0.0$ $f = 100.0$

(Note that $x = x_u$ and $y = x_l$)

Table 3 Median function evaluations for the upper level (UL) and the lower level (LL) from 31 runs of different algorithms

	UL func. evals.			LL func. evals.			Savings	
	φ-appx Med	Ψ-appx Med	No-appx Med	φ-appx Med	Ψ-appx Med	No-appx Med	φ	Ψ
TP1	134	150	–	1438	2061	–	Large	Large
TP2	148	193	436	1498	2852	5686	73 %	50 %
TP3	187	137	633	2478	1422	6867	64 %	79 %
TP4	299	426	1755	3288	6256	19764	83 %	69 %
TP5	175	270	576	2591	2880	6558	61 %	56 %
TP6	110	94	144	1489	1155	1984	25 %	41 %
TP7	166	133	193	2171	1481	2870	24 %	47 %
TP8	212	343	403	2366	5035	7996	69 %	36 %

The savings represent the proportion of total function evaluations (LL+UL) saved because of using the approximation when compared with no-approximation approach

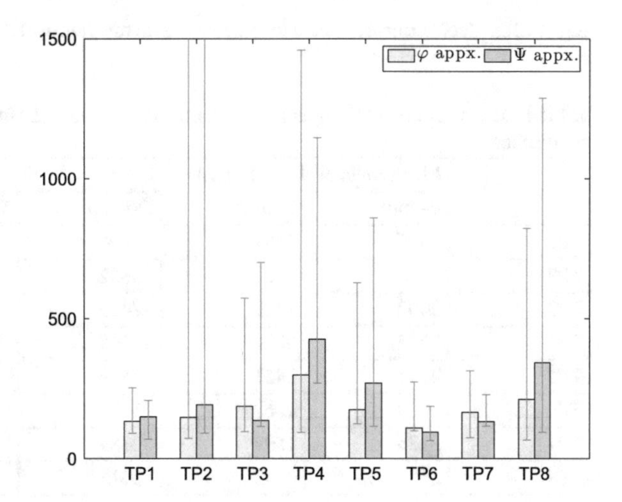

Fig. 7 Box plot (31 runs/samples) for the upper level function evaluations required for test problems 1–8

the problems as compared to the no-approximation approach. The savings column in the table shows the proportion of function evaluations savings that can be directly attributed to φ and Ψ approximations. Slight difference in performance between the two approximation strategies can be attributed to the quality of approximations achieved for specific test problems. To provide the readers an idea about the extent of savings in function evaluations obtained from using metamodeling based strategies, we also provide comparisons with earlier evolutionary approaches [43, 44] in Table 4. These approaches are based on single-level reduction using lower level KKT conditions. A significantly poor performance of these methods emphasizes the fact that even when it is possible to write the KKT constraints for the lower level problem, a single level reduction might not necessarily make the problem easy to solve.

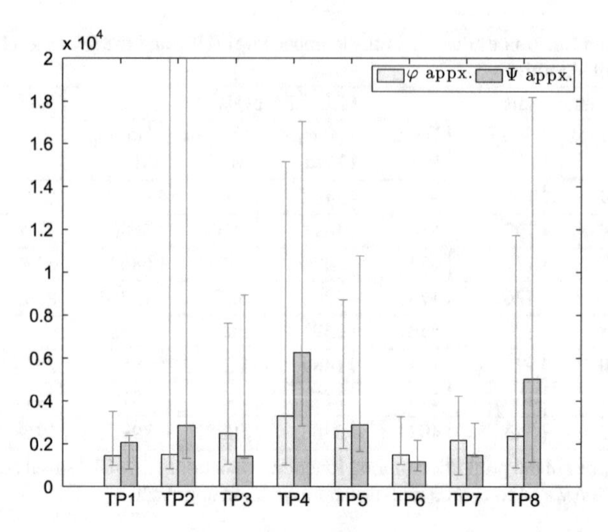

Fig. 8 Box plot (31 runs/samples) for the lower level function evaluations required for test problems 1–8

Table 4 Mean of the sum of upper level (UL) and lower level (LL) function evaluations for different approaches

	Mean func. evals. (UL+LL)				
	φ-appx.	Ψ-appx.	No-appx.	WJL [43]	WLD [44]
TP1	1595	2381	35896	85499	86067
TP2	1716	3284	5832	256227	171346
TP3	2902	1489	7469	92526	95851
TP4	3773	6806	21745	291817	211937
TP5	2941	3451	7559	77302	69471
TP6	1689	1162	1485	163701	65942
TP7	2126	1597	2389	1074742	944105
TP8	2699	4892	5215	213522	182121

It is noteworthy that the Ψ-mapping in a bilevel optimization problem could be a set-valued mapping as shown in Fig. 9, i.e. for some or all upper level decision vectors in the search space, the lower level optimization problem may have multiple optimal solutions. Such a situation offers dual challenges; first, finding the Ψ-set is difficult; second, approximating the set is also difficult. In such cases approximating the Ψ-mapping will not help. To test this hypothesis, we modified all the 8 test problems by adding two additional lower level variables (y_p and y_q) that makes the Ψ-mapping in all the test problems as set-valued for the entire domain of Ψ. The modification

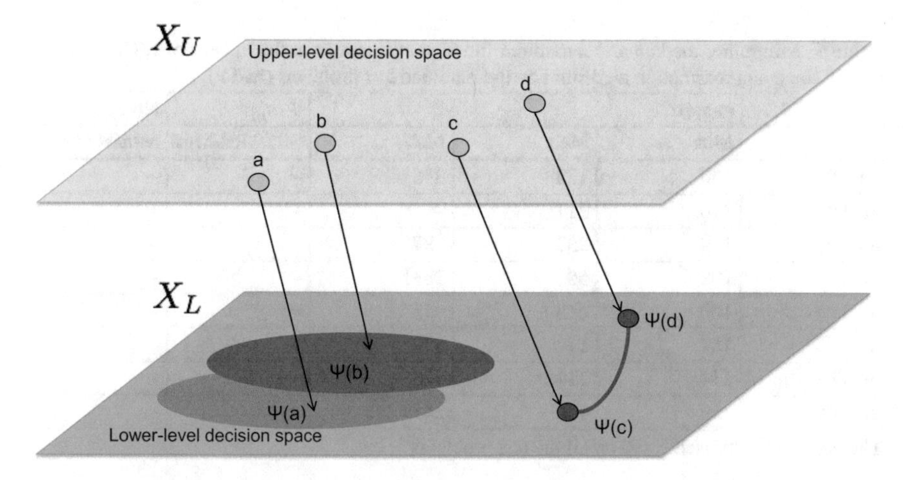

Fig. 9 A scenario where the Ψ-mapping is set-valued in some regions and single-valued in other regions

does not change the original bilevel solution. This was achieved by modifying the upper and lower level functions for all the test problems as follows:

$$F^{new}(x, y) = F(x, y) + y_p^2 + y_q^2$$
$$f^{new}(x, y) = f(x, y) + (y_p - y_q)^2$$
$$y_p, y_q \in [-1, 1]$$

Note that the above modification necessarily makes the lower level problem have multiple optimal solutions corresponding to all x, as the added term gets minimized at $y_p = y_q$ which has infinitely many solutions. Out of the infinitely many lower level optimal solutions, the upper level prefers $y_p = y_q = 0$. With this simple modification, we execute our algorithm with φ-approximation and Ψ-approximation on all test problems, the results for which are presented through Tables 5 and 6. For all the problems, the Ψ-approximation idea fails. The φ-approximation idea continues to work effectively as before. The slight increase in function evaluations for the Ψ-approximation approach comes from the fact that there are additional variables in the problem.

To conclude, the Ψ-mapping offers the advantage that if it can be approximated accurately, it readily gives the optimal lower level variables. However, in cases when this mapping is set-valued, approximating Ψ can be very difficult. On the other hand, the φ-mapping is always single-valued, approximating which is much easier, and is therefore more preferred over the Ψ-mapping. The results shown in this section clearly demonstrate that even a simple modification that leads to multiple lower level optimal solutions, makes the Ψ-approximation strategy fail because of poor quality of approximation. To our best knowledge, most of the studies utilizing metamodeling techniques to solve bilevel optimization problems have mostly relied on approximat-

Table 5 Minimum, median and maximum function evaluations at the upper level (UL) from 31 runs of the φ-approximation algorithm on the modified test problems (m-TP)

	φ-appx.			Ψ-appx.	No-appx.
	Min	Med	Max	Min/Med/Max	Min/Med/Max
m-TP1	130	172	338	–	–
m-TP2	116	217	–	–	–
m-TP3	129	233	787	–	–
m-TP4	198	564	2831	–	–
m-TP5	160	218	953	–	–
m-TP6	167	174	529	–	–
m-TP7	114	214	473	–	–
m-TP8	150	466	2459	–	–

The other two approaches fail on all the test problems

Table 6 Minimum, median and maximum function evaluations at the lower level (LL) from 31 runs of the φ-approximation algorithm on the modified test problems (m-TP)

	φ-appx.			Ψ-appx.	No-appx.
	Min	Med	Max	Min/Med/Max	Min/Med/Max
m-TP1	2096	2680	8629	–	–
m-TP2	2574	4360	–	–	–
m-TP3	1394	3280	13031	–	–
m-TP4	1978	5792	28687	–	–
m-TP5	3206	4360	17407	–	–
m-TP6	2617	3520	8698	–	–
m-TP7	1514	5590	11811	–	–
m-TP8	2521	6240	35993	–	–

The other two approaches fail on all the test problems

ing the Ψ-mapping. Given the ease and reliability offered by the φ-approximation over Ψ-approximation, we believe that future research on metamodeling-based techniques should closely look at the benefits of the φ-approximation.

5 Multiobjective Bilevel Optimization

A substantial body of research exists on single-objective bilevel optimization, but relatively few papers have considered bilevel problems with multiple objectives on both levels. Even less research has been done to understand the impacts of decision-interaction and uncertainty that arise in multiobjective bilevel problems. One of the reasons for little research in the area is that the problem becomes both mathematically and computationally intractable even with simplifying assumptions like continuity, differentiability, convexity etc. However, given that multiobjective bilevel problems exist in practice, researchers have tried to explore ideas to handle these problems.

Some of the studies on multiobjective bilevel optimization that exist are mostly directed towards development of techniques for solving optimistic formulation of the problem, where the decision-makers are assumed to co-operate and the leader can freely choose any Pareto-optimal lower-level solution. Studies by Eichfelder [57, 58] utilize classical techniques to solve simple multiobjective bilevel problems. The lower level problems are handled using a numerical optimization technique, and the upper level problem is handled using an adaptive exhaustive search method. This makes the solution procedure computationally demanding and non-scalable to large-scale problems. The method is close to a nested strategy, where each of the lower level optimization problems is solved to Pareto-optimality. Shi and Xia [59] use the ϵ-constraint method at both levels of a multiobjective bilevel problem to convert the problem into an ϵ-constraint bilevel problem. The ϵ-parameter is elicited from the decision maker, and the problem is solved by replacing the lower level constrained optimization problem with its KKT conditions. The problem is solved for different ϵ-parameters, until a satisfactory solution is found.

With the surge in computation power, a number of nested evolutionary algorithms have also been proposed, which solve the lower level problem completely for every upper level vector to arrive at the problem optima. One of the first studies, utilizing an evolutionary approach for bilevel multiobjective algorithms was in [35]. The study involved multiple objectives at the upper level, and a single objective at the lower level. The study suggested a nested genetic algorithm, and applied it on a transportation planning and management problem. Later [60] used a particle swarm optimization (PSO)-based nested strategy to solve a multi-component chemical system. The lower level problem in their application problem was linear for which they used a specialized linear multiobjective PSO approach. Recently, a hybrid bilevel evolutionary multiobjective optimization algorithm approach coupled with local search was proposed in [61]. In the paper, the authors handled nonlinear as well as discrete bilevel problems with a relatively large number of variables. The study also provided a suite of test problems for bilevel multiobjective optimization. An extension to this study [62] attempted to solve bilevel multiobjective optimization with fewer function evaluations by interacting with the leader. The idea in this study was to interact with the upper level decision maker only to model her preferences and find the most preferred Pareto-optimal point instead of the entire frontier. The study borrowed ideas from the area of preference-based evolutionary algorithms.

Until recently, the focus has been primarily on algorithms for handling deterministic problems. Less emphasis has been paid to the decision-making intricacies that arise in practical multiobjective bilevel problems. The first concern is the reliance on the assumption that transfers decision-making power to the leader by allowing her to freely choose any Pareto-optimal solution from the lower-level optimal frontier. In practical problems, the preferences of the lower-level decision maker may not be aligned with the leader. Although a leader can anticipate the follower's actions and optimize her strategy accordingly, it is unrealistic to assume that she can decide which trade-off the follower should choose. To solve hierarchical problems with conflicting decision-makers, a few studies have proposed a line of interactive fuzzy programming models [63, 64]. The methods have been successfully used to handle decentralized bilevel problems that have more than one lower level decision maker [65]. However, the assumption of mutual co-operation and repeated interactions between decision-makers is not necessarily feasible; e.g., in homeland security applications and competitive business decisions. The second concern is the decision-uncertainty. The strategy chosen by the follower may well deviate from what is expected by the leader, which thus gives rise to uncertainty about the realized outcome. It is worthwhile to note that the notion of decision-uncertainty that emanates from not knowing the follower's preferences exactly is different from the uncertainty that follows from non-preference related factors such as stochastic model parameters.

6 Multiobjective Bilevel Optimization and Decision Making

In this section, we provide three different formulations for a multiobjective bilevel optimization problem. First, we consider the standard formulation, where there is no decision making involved at the lower level and all the lower level Pareto-optimal solutions are considered at the upper level (see Fig. 10). Second, we consider a formulation, where the decision maker acts at the lower level and chooses a solution to her liking. The preference structure of the follower is known to the leader and can

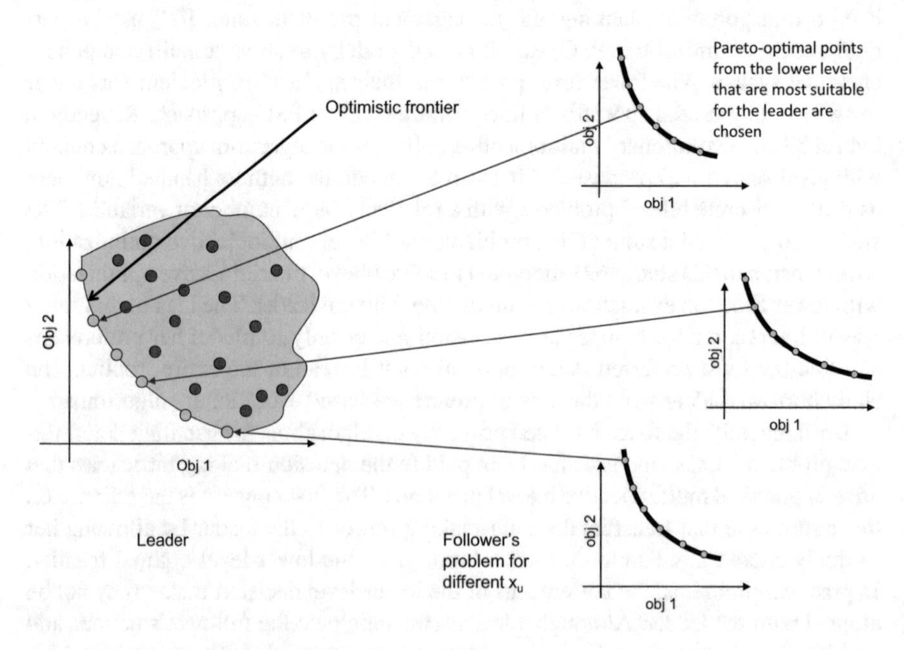

Fig. 10 Optimistic bilevel multiobjective optimization

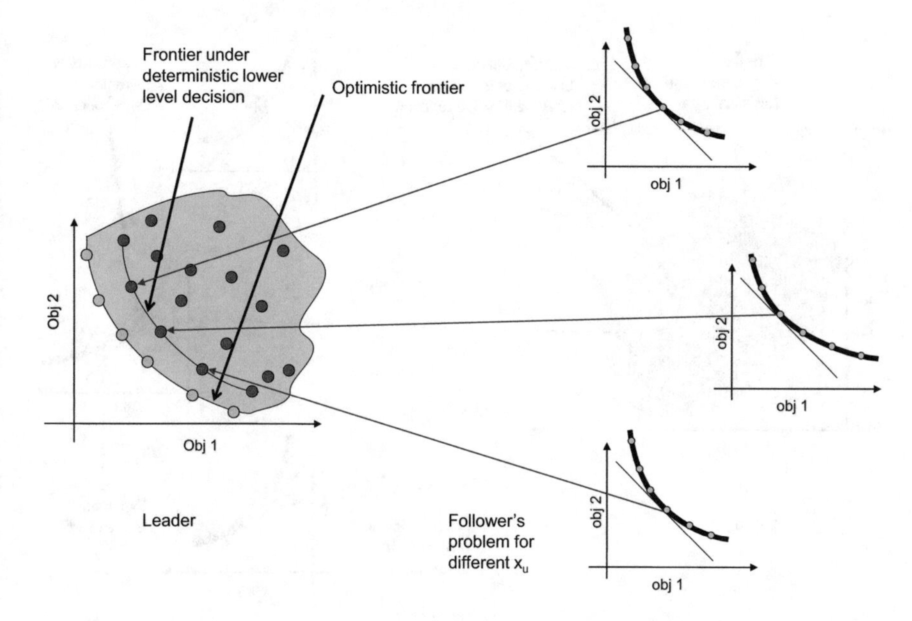

Fig. 11 Bilevel multiobjective optimization with deterministic lower level decisions

be modeled as a value function (see Fig. 11). Finally, we discuss a problem, where the lower level decision maker's preferences are not known with certainty and the upper level decision maker needs to take this decision-uncertainty into account when choosing her optimal strategy (see Fig. 12).

6.1 Multiobjective Bilevel Optimization: The Optimistic Formulation

Bilevel multiobjective optimization is a nested optimization problem involving two levels of multiobjective optimization tasks. The structure of a bilevel multiobjective problem demands that only the Pareto-optimal solutions to the lower level optimization problem may be considered as feasible solutions for the upper level optimization problem. There are two classes of variables in a bilevel optimization problem; namely, the upper level variables $x_u \in X_U \subset \mathbb{R}^n$, and the lower level variables $x_l \in X_L \subset \mathbb{R}^m$. The lower level multiobjective problem is solved with respect to the lower level variables, x_l, and the upper level variables, x_u act as parameters to the optimization problem. Each x_u corresponds to a different lower level optimization problem, leading to a different Pareto-optimal front. The upper level problem is optimized with respect to both classes of variables, $x = (x_u, x_l)$.

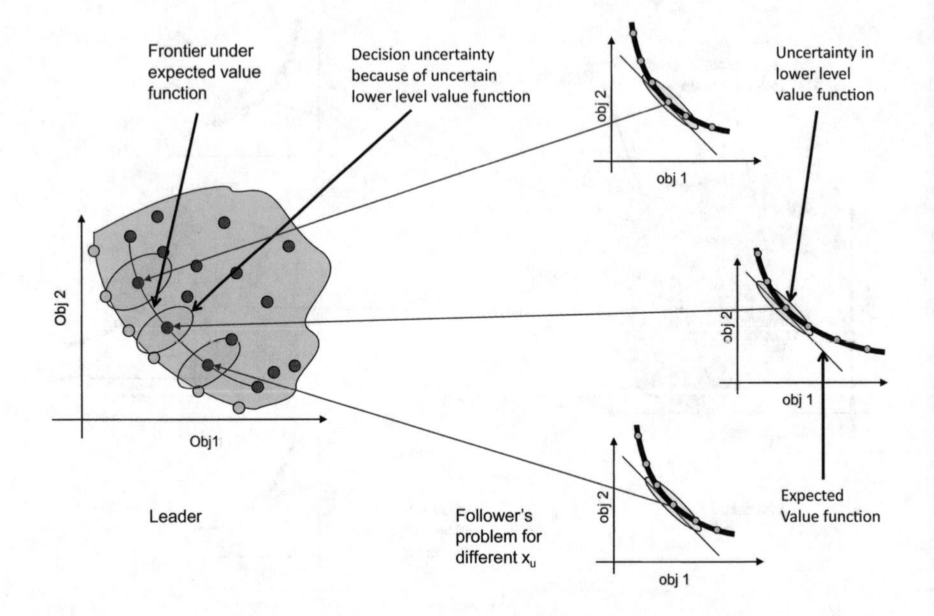

Fig. 12 Bilevel multiobjective optimization with uncertainty in lower level decisions

Definition 5 For the upper-level objective function $F : \mathbb{R}^n \times \mathbb{R}^m \to \mathbb{R}^p$ and lower-level objective function $f : \mathbb{R}^n \times \mathbb{R}^m \to \mathbb{R}^q$, the bilevel problem is given by

$$\begin{aligned}
&\min_{x_u \in X_U, x_l \in X_L} \quad F(x_u, x_l) = (F_1(x_u, x_l), \ldots, F_p(x_u, x_l)) \\
&\text{subject to} \\
&\quad x_l \in \underset{x_l}{\operatorname{argmin}}\{f(x_u, x_l) = (f_1(x_u, x_l), \ldots, f_q(x_u, x_l)) : \\
&\qquad\qquad g_j(x_u, x_l) \leq 0, j = 1, \ldots, J\} \\
&\quad G_k(x_u, x_l) \leq 0, k = 1, \ldots, K
\end{aligned}$$

where $G_k : X_U \times X_L \to \mathbb{R}$, $k = 1, \ldots, K$ denote the upper level constraints, and $g_j : X_U \times X_L \to \mathbb{R}$ represent the lower level constraints, respectively. Equality constraints may also exist that have been avoided for brevity.

An equivalent formulation of the above problem can be stated in terms of set-valued mappings as follows:

Definition 6 Let $\Psi : \mathbb{R}^n \rightrightarrows \mathbb{R}^m$ be a set-valued mapping,

$$\begin{aligned}
\Psi(x_u) = \underset{x_l}{\operatorname{argmin}}\{&f(x_u, x_l) = (f_1(x_u, x_l), \ldots, f_2(x_u, x_l)) : \\
&g_j(x_u, x_l) \leq 0, j = 1, \ldots, J\},
\end{aligned}$$

which represents the constraint defined by the lower-level optimization problem, i.e. $\Psi(x_u) \subset X_L$ for every $x_u \in X_U$. Then the bilevel multiobjective optimization problem can be expressed as a constrained multiobjective optimization problem:

$$\begin{aligned} \min_{x_u \in X_U, x_l \in X_L} \quad & F(x_u, x_l) = (F_1(x_u, x_l), \ldots, F_p(x_u, x_l)) \\ \text{subject to} \quad & x_l \in \Psi(x_u) \\ & G_k(x_u, x_l) \leq 0, k = 1, \ldots, K \end{aligned}$$

where Ψ can be interpreted as a parameterized range-constraint for the lower-level decision vector x_l.

In the above two formulations, the lower level decision maker is assumed to cooperate with the upper level decision maker, such that she provides all Pareto-optimal points to the upper level decision maker who then chooses the best point according to the upper level objectives. The assumption effectively reduces the influence of the follower and transfers the decision-making power to the leader. Alternatively, one can say that the lower-level decision maker is assumed to be indifferent to all lower-level Pareto-optimal solutions. Though this formulation has been studied in the past, it is a highly unrealistic formulation where decision making aspects at the lower level are not taken into account.

Next, we demonstrate the optimistic formulation through a simple multiobjective bilevel optimization problem taken from [58].

Example 1 The problem has a single upper level and two lower level variables; such that $x_u = (x)$ and $x_l = (y_1, y_2)^T$. The formulation of the problem is given below:

$$\begin{aligned} & \text{Minimize } F(x, y_1, y_2) = \begin{pmatrix} y_1 - x \\ y_2 \end{pmatrix}, \\ & \text{subject to } (y_1, y_2) \in \underset{(y_1, y_2)}{\text{argmin}} \left\{ f(x, y_1, y_2) = \begin{pmatrix} y_1 \\ y_2 \end{pmatrix} \middle| g_1(x) = x^2 - y_1^2 - y_2^2 \geq 0 \right\}, \\ & \quad G_1(x) = 1 + y_1 + y_2 \geq 0, \\ & \quad -1 \leq y_1, y_2 \leq 1, \quad 0 \leq x \leq 1. \end{aligned} \tag{1}$$

The Pareto-optimal set for the lower level optimization task for a given x is the bottom-left quarter of the circle: $\{(y_1, y_2) \in \mathbb{R}^2 \mid y_1^2 + y_2^2 = x^2, y_1 \leq 0, y_2 \leq 0\}$. Lower level frontiers corresponding to different x are shown in Fig. 13. As observed from the figure, the linear constraint at the upper level does not allow the entire quarter circle to be feasible for some x. Therefore, at most two points from the quarter circle belong to the upper level Pareto-optimal set of the bilevel problem that is shown in Fig. 14. The lower level frontiers for different x are also plotted in the upper level objective space. Figures 13 and 14 also show three points A, B and C for $x = 0.9$, where points A and B participate in the upper level frontier while point C is rendered infeasible because of the upper level constraint. The analytical Pareto-optimal set for this problem is given as:

$$\left\{ (x, y_1, y_2) \in \mathbb{R}^3 \;\middle|\; x \in \left[\frac{1}{\sqrt{2}}, 1\right], y_1 = -1 - y_2, y_2 = -\frac{1}{2} \pm \frac{1}{4}\sqrt{8x^2 - 4} \right\}. \tag{2}$$

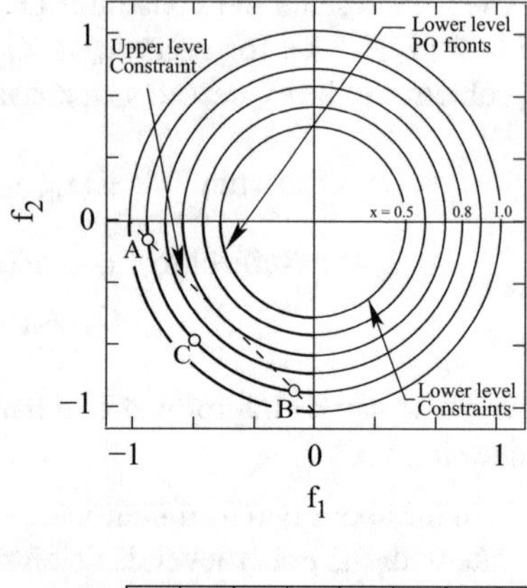

Fig. 13 Lower level Pareto-optimal fronts for different x_u in lower level objective space

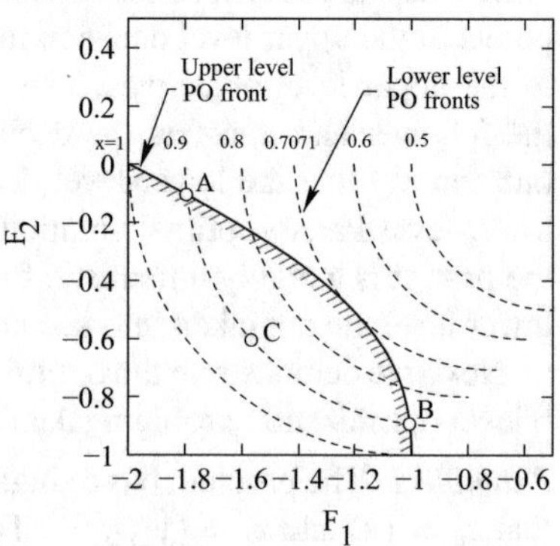

Fig. 14 Upper level Pareto-optimal front and lower level fronts in upper level objective space

This problem demonstrates that leader takes all the lower level Pareto-optimal solutions and then based on her constraints and non-domination criterion decides the solutions to be kept. Once this multiobjective bilevel problem is given, the upper level Pareto-frontier can be identified without considering any decision making aspects.

Some of the studies that attempted to handle the optimistic formulation are [57, 58] in the area of mathematical optimization and [61, 66] in the area of evolutionary computation. In [61], authors utilize a hierarchical evolutionary multiobjective optimization approach to solve a number of difficult multiobjective bilevel problems. Though the approach retains a nested structure, a number of intelligent schemes were employed that led to savings, when compared to a brute force nested algorithm. Some of the ideas utilized include; adjusting the number of subpopulations and their sizes for lower level search adaptively, solving the lower level problem with an evolutionary algorithm for a few generations and then employing local search on members that are likely to participate in upper level non-dominated frontier, utilizing a hypervolume-based termination criterion at both levels, and using archive that keeps those solutions that are feasible (with respect to constraints and lower level problem) and non-dominated at the upper level.

Table 7 Ratio of median function evaluations required by Algorithm-1 [61] against Algorithm-3 [66] and Algorithm-2 (purely nested) against Algorithm-3 [66]

Pr. No.	$\frac{Algorithm-1}{Algorithm-3}$		$\frac{Algorithm-2}{Algorithm-3}$	
	Total LL FE	Total UL FE	Total LL FE	Total UL FE
DS1	1.54	1.23	17.51	13.58
DS2	1.33	1.11	17.07	11.33
DS3	1.43	1.19	18.03	11.21
DS4	1.28	1.25	16.06	13.59
DS5	1.32	1.21	19.89	12.27

Table 8 Function evaluations (FE) required by Algorithm-3 for the upper level (UL) and lower level (LL)

Pr. No. (var.)	Best		Median		Worst	
	Total LL FE	Total UL FE	Total LL FE	Total UL FE	Total LL FE	Total UL FE
DS1 (20)	1946496	72334	2215966	74502	2430513	86697
DS2 (20)	3728378	93015	3728256	110006	4584177	126416
DS3 (20)	2540181	90754	3295798	100015	3733238	104025
DS4 (10)	904806	33804	1118631	42986	1339842	50686
DS5 (10)	1187359	38477	1356863	47071	1684170	59325

Best, median and worst values have been computed from 21 runs of the algorithm on each test problem. The lower level function evaluations include the evaluations of local search as well

Recently, along the lines of Ψ-mapping approximation, a multi-fiber approach has been proposed in [66]. In this approach the authors attempt to approximate the Ψ-mapping using multiple discrete fibers. The Ψ-mapping is more likely to be a (moving) set in the context of multiobjective bilevel optimization; therefore, ideas that can approximate sets have to be employed. This is one of the approaches the tries to exploit the structure and properties of the problem to solve it. The scheme can not be termed nested, but still requires solving some instances of the lower level problem to construct an approximation of the Ψ-mapping. In Table 7 we provide the results for three algorithms; algorithm 1 [61], algorithm 2 (purely nested) and algorithm 3 [66]; on a set of 5 test problems [61, 67]. The numbers in the table represent the ratio of function evaluations required by algorithm 1 and algorithm 2 with respect to algorithm 3. The function evaluations for algorithm 3 can be found in Table 8.

Before concluding the discussion on the optimistic formulations and solution procedures for multiobjective bilevel optimization, we would like to highlight that it is possible to write this formulation with multiple objectives at upper level and single objective at lower level. However, this comes at the cost of increased variables at the upper level. The following formulation has been known in mathematical optimization, but one of the first studies in the context of evolutionary optimization can be found in [68].

Definition 7 For a scalarizing function $S : \mathbb{R}^p \times \mathbb{R}^p \rightarrow \mathbb{R}$ with weight vector $w \in W \subset \mathbb{R}^p$

$$\begin{aligned} \min_{x_u \in X_U, x_l \in X_L, w \in W} \quad & F(x_u, x_l) \\ \text{subject to} \quad & x_l \in \operatorname*{argmin}_{x_l \in X_L} \{ S(f(x_u, x_l), w) \\ & \qquad\qquad \text{subject to} \quad g_j(x_u, x_l) \leq 0, j = 1, \ldots, J \} \\ & G_k(x_u, x_l) \leq 0, k = 1, \ldots, K, \end{aligned}$$

where w acts as an upper level vector along with x_u.

It is important in the above formulation that the scalarizing function is able to span the entire lower level Pareto-optimal set through different values of w. The idea behind the formulation is that by changing w, one can select different Pareto-optimal solutions from the lower level corresponding to each upper level decision vector.

6.2 *Multiobjective Bilevel Optimization with Deterministic Decisions at Lower Level*

Considering the decision-making situations that arise in practice, a departure from the assumption of an indifferent lower level decision maker is necessary. Rather than providing all Pareto-optimal points to the leader, the follower is likely to act according to her own interests and choose the most preferred lower level solution herself. As a result, the allowance of lower level decision making has a substantial impact on the formulation of multiobjective bilevel optimization problems. First, the lower level problem can no longer be viewed as a range-constraint that depends only on lower-level objectives. Instead it is better interpreted as a selection function that maps a given upper level decision to a corresponding Pareto-optimal lower level solution that is most preferred by the follower. Second, in order to solve the bilevel problem, the upper level decision maker now needs to model the follower's behavior by anticipating her preferences towards different objectives. Naturally, these changes lead to a number of intricacies that were not encountered in the previous formulations. This formulation assumes perfect information to the leader about the follower's preference structure. Using the preference structure information it is possible to reduce the lower level problem into a single objective optimization problem [69].

Definition 8 Let $\xi \in \Xi$ denote a vector of parameters describing the follower's preferences. If the upper level decision maker has complete knowledge of the follower's preferences, the follower's actions can then be modeled via selection mapping

$$\sigma : X_U \times \Xi \rightarrow X_L, \quad \sigma(x_u, \xi) \in \Psi(x_u), \tag{3}$$

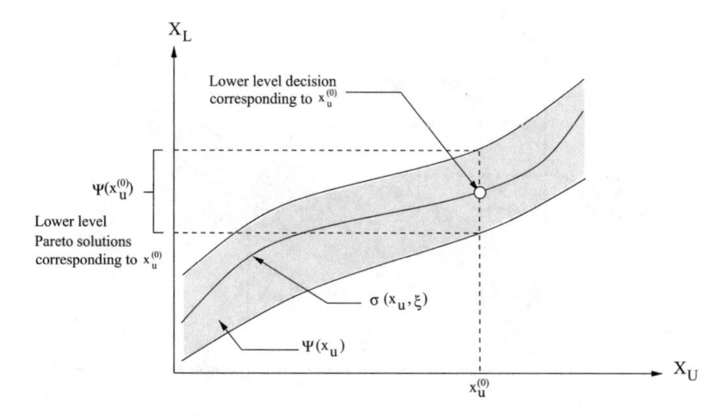

Fig. 15 Decision making under certainty

where Ψ is the set-valued mapping given by Definition 2. The resulting bilevel problem can be rewritten as follows:

$$\begin{aligned} \min_{x_u \in X_U} \quad & F(x_u, x_l) = (F_1(x_u, x_l), \ldots, F_p(x_u, x_l)) \\ \text{subject to} \quad & x_l = \sigma(x_u, \xi) \in \Psi(x_u) \\ & G_k(x_u, x_l) \leq 0, k = 1, \ldots, K \end{aligned} \tag{4}$$

To illustrate the definition, consider Fig. 15, where the shaded region

$$\text{gph}\, \Psi = \{(x_u, x_l) \ : \ x_l \in \Psi(x_u)\} \tag{5}$$

represents the follower's Pareto-optimal solutions $\Psi(x_u)$ for any given leader's decision x_u. These are the rational reactions, which the follower may choose depending on her preferences. If the leader is aware of the follower's objectives, she will be able to identify the shaded region completely by solving the follower's multiobjective optimization problem for all x_u. However, if the follower is able to act according to her own preferences, she will choose only one preferred solution $\sigma(x_u, \xi)$ for every upper level decision x_u. When the preferences of the follower are perfectly known, the leader can identify $\sigma(\cdot, \xi)$ that characterizes follower's rational reactions for different x_u, and solve the hierarchical optimization task completely.

6.3 *Multiobjective Bilevel Optimization with Lower Level Decision Uncertainty*

The assumption that the follower's preferences are perfectly known to the leader itself might be an inaccurate description of real life scenarios. Most practitioners

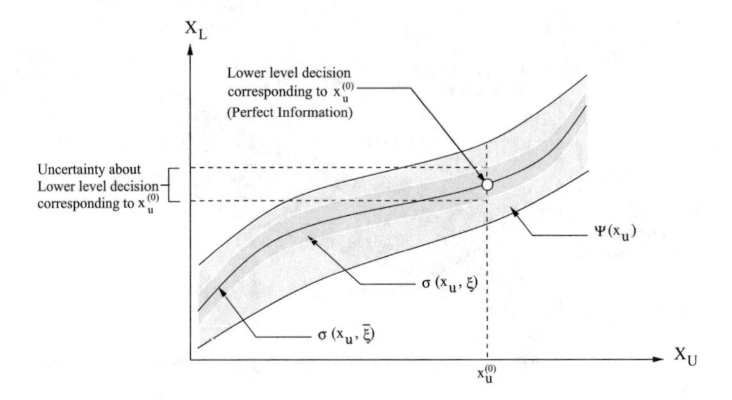

Fig. 16 Decision making under uncertainty

would find it hard to accept this even when constructing approximations. A natural path towards a more realistic framework would be to relax the axiom of perfect information by assuming that the leader is only partially aware of the follower's preferences. This lack of information leads to the notion of lower level decision uncertainty that is experienced by the leader while solving the bilevel optimization task [70].

For illustration, consider Fig. 16, where the expected behavior of the follower is shown as the graph of the selection mapping $\sigma(\cdot, \bar{\xi})$, where $\bar{\xi}$ represents the expected preference known to the leader. The narrow dark shaded band shows the region of uncertainty in which the follower makes her decisions. For different preferences ξ, $\sigma(\cdot, \xi)$ represents the corresponding decisions of the follower. If the leader is aware of the follower's objectives, the uncertainty region identified by a random ξ is always bounded by gph Ψ because $\sigma(x_u, \xi) \in \Psi(x_u)$ for all $x_u \in X_U$ and $\xi \in \Xi$. However, it is noteworthy that this band is not directly available to the leader but needs to be modeled. In a situation, where the leader cannot elicit follower's preferences by interacting with the follower, a feasible strategy is to utilize the prior information she has about the follower and incorporate it in a tractable stochastic model that characterizes the follower's behavior.

To accommodate the decision uncertainty, we assume that the follower's preferences are described by a random variable $\xi \sim \mathcal{D}_\xi$, which takes values in a set Ξ of $\mathbb{R}^q$. The probability distribution $\mathcal{D}_\xi$ reflects the leader's uncertainty and prior information about follower's expected behavior. In this framework, the assumption of preference uncertainty is equivalent to saying that the lower level decision is a random variable with a distribution that is parametrized by a given upper level decision x_u, i.e. $x_l \sim \mathcal{D}_\sigma(x_u)$. This means that the lower level decision uncertainty experienced by the leader will vary point-wise depending on the follower's objectives and the leader's own decision.

For demonstration of the uncertainty aspects in the objective spaces of the leader and the follower, consider Figs. 17 and 18 that show two different scenarios. In the first scenario, we assume a deterministic situation where the follower's prefer-

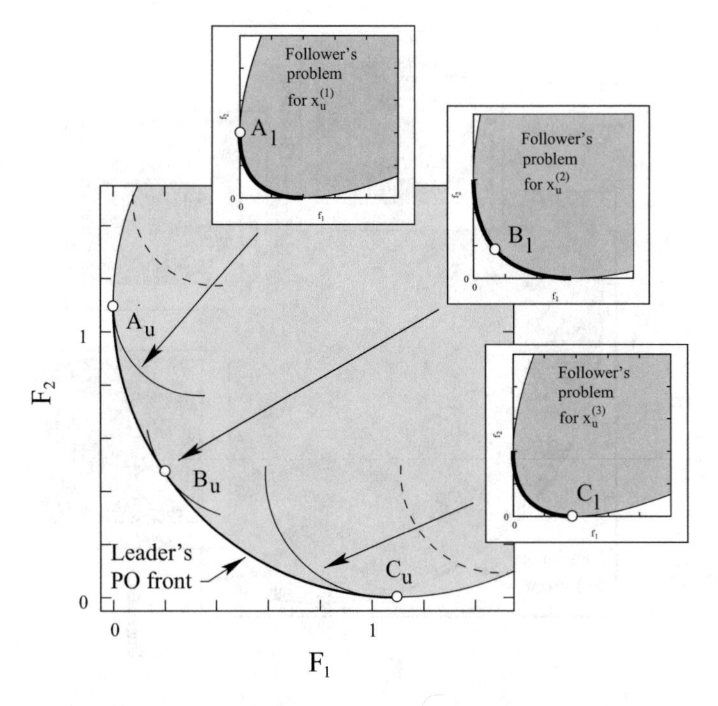

Fig. 17 Insets: follower's problem for different x_u. A_l, B_l and C_l represent the follower's decisions for $x_u^{(1)}$, $x_u^{(2)}$ and $x_u^{(3)}$ respectively. A_u, B_u and C_u are the corresponding points for the leader in the leader's objective space

ences and actions are known with certainty. Both leader and follower are assumed to have two objectives, i.e., $p = q = 2$. In this case, the leader solves the bilevel problem in Definition 8 under perfect information. Therefore, each point on the leader's Pareto-frontier corresponds to one of the points on the follower's Pareto-frontier. If $\bar{\xi}$ is the given vector of follower's preferences, then for any leader's choice $x_u^{(i)}$ the corresponding lower level decision is given by $x_l^{(i)} = \sigma(x_u^{(i)}, \bar{\xi})$. This is shown in Fig. 17, where the upper level points $A_u = F(x_u^{(1)}, \sigma(x_u^{(1)}, \bar{\xi}))$, $B_u = F(x_u^{(2)}, \sigma(x_u^{(2)}, \bar{\xi}))$, and $C_u = F(x_u^{(3)}, \sigma(x_u^{(3)}, \bar{\xi}))$ are paired with the points $A_l = f(x_u^{(1)}, \sigma(x_u^{(1)}, \bar{\xi}))$, $B_l = f(x_u^{(2)}, \sigma(x_u^{(2)}, \bar{\xi}))$ and $C_l = f(x_u^{(3)}, \sigma(x_u^{(3)}, \bar{\xi}))$ that lie on the follower's Pareto-front for $x_u^{(1)}$, $x_u^{(2)}$, and $x_u^{(3)}$, respectively.

The situation can be contrasted from another scenario shown in Fig. 18, where the follower's preferences are uncertain. The leader is still assumed to be fully aware of the form of σ, but she no longer knows the true value of ξ. By assuming a prior information $\xi \sim \mathcal{D}_\xi$, the leader can attempt to solve the bilevel problem based on the expected preferences of the follower, i.e.

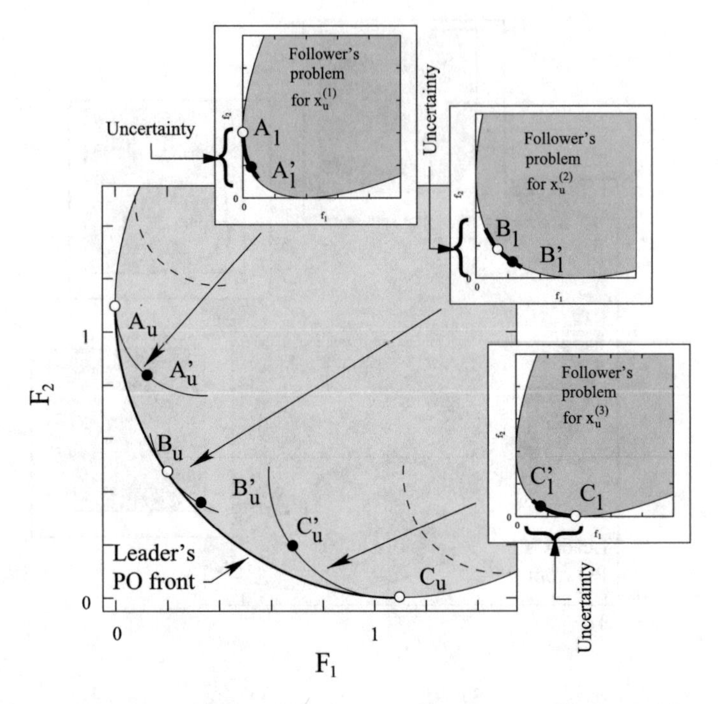

Fig. 18 Insets: follower's problem for different x_u. A_l, B_l and C_l are the expected decisions of the follower. $A_l^{'}$, $B_l^{'}$ and $C_l^{'}$ are the actual decisions that the follower takes. The corresponding points for the leader are shown in the leader's objective space

$$\begin{aligned} \min_{x_u \in X_U} \quad & F(x_u, \bar{x}_l) \\ \text{subject to} \quad & \bar{x}_l = \sigma(x_u, E[\xi]) \in \Psi(x_u), \quad \xi \sim \mathcal{D}_\xi \\ & G_k(x_u, \bar{x}_l) \leq 0, k = 1, \ldots, K. \end{aligned} \tag{6}$$

For convenience of the example, we assume that the expected actions are the same as the actions in Fig. 17, i.e., $\sigma(x_u, E[\xi])] = \sigma(x_u, \bar{\xi})$ for all x_u. As a result, the leader obtains a *Pareto-frontier corresponding to the follower's expected value function* (POF-EVF). However, when she begins to implement the given strategies, the follower's realized actions may deviate from the expected strategies obtained by solving (6). Since ξ is uncertain from the leader's perspective, the follower's true preferences ξ can differ from $\bar{\xi}$ that was expected based on prior information. As shown in the figure, for any strategy $x_u^{(1)}$, $x_u^{(2)}$ or $x_u^{(3)}$ chosen by the leader, the follower may prefer to choose $A_l^{'}$, $B_l^{'}$ or $C_l^{'}$ instead of A_l, B_l or C_l expected by the leader. It is found that because of the follower's deviation from the expected actions, the leader no longer operates on the POF-EVF. In the objective space, the uncertainty experienced by the leader is reflected in the probability and size of deviations away from the POF-EVF. The follower, on the other hand, does not experience similar uncertainty, because she can always observe the action taken by the leader before making her own decision.

Depending on the problem, uncertainty of the lower level decision maker's preferences may lead to significant losses at the upper level. Therefore, the leader would like to solve the bilevel problem taking the uncertainties into account. While making a decision, the leader might prefer those regions on its frontier, which are less sensitive to lower level uncertainties and at the same time offer an acceptable trade-off between the objectives. For instance, in the context of the above example, we observe that the expected variation in the objective space is considerably less at the region corresponding to $x_u^{(2)}$ than at $x_u^{(1)}$ or $x_u^{(3)}$. If the leader chooses this point, she knows that the realized upper level objective values are only little affected by the actions of the lower level decision maker. From the perspective of practical decision making, it is valuable for the leader to be aware of the level of uncertainty associated with different strategies.

7 Future Research Directions

In this chapter, we have tried to provide an introduction to the work done in the area of bilevel optimization using evolutionary algorithms. The main topics covered include;

1. Single objective bilevel optimization and promising ideas that might be useful in solving complex bilevel problems.
2. Multiobjective bilevel optimization methods and decision making intricacies.

While the above two topics themselves offer significant opportunity of future research, there also exist other areas within bilevel optimization that are less explored and offer potential for future research. For instance, there can be other forms of uncertainties in bilevel optimization, like, variable and parameter uncertainties. Some preliminary work on these topics can be found in [71, 72]. With an increase in computational power, there is an enormous scope of development of distributed computing methods that can solve bilevel problems with large number of variables or objectives in a short time. However, at this point it is worth mentioning that in the last decade a number of evolutionary algorithms have been developed that are computationally very expensive and purely nested. Future research ideas on evolutionary computation should rely also on exploiting the structure and properties of bilevel problems, which will ensure better scalability of the procedures. To conclude, almost every other discipline faces application problems that are bilevel in nature. This offers application oriented research opportunities both from modeling and solution perspectives.

References

1. Von Stackelberg, H.: The Theory of the Market Economy. Oxford University Press, New York (1952)
2. Bracken, J., McGill, J.T.: Mathematical programs with optimization problems in the constraints. Op. Res. **21**(1), 37–44 (1973)
3. Wen, U.-P., Hsu, S.-T.: Linear bi-level programming problems–a review. J. Op. Res. Soc. 125–133 (1991)
4. Ben-Ayed, O.: Bilevel linear programming. Comput. Op. Res. **20**(5), 485–501 (1993)
5. Bard, J.F., Moore, J.T.: A branch and bound algorithm for the bilevel programming problem. SIAM J. Sci. Stat. Comput. **11**(2), 281–292 (1990)
6. Edmunds, T.A., Bard, J.F.: Algorithms for nonlinear bilevel mathematical programs. IEEE Trans. Syst. Man Cybern. **21**(1), 83–89 (1991)
7. Al-Khayyal, F.A., Horst, R., Pardalos, P.M.: Global optimization of concave functions subject to quadratic constraints: an application in nonlinear bilevel programming. Ann. Op. Res. **34**(1), 125–147 (1992)
8. Liu, G., Han, J., Wang, S.: A trust region algorithm for bilevel programing problems. Chin. Sci. Bull. **43**(10), 820–824 (1998)
9. Hansen, P., Jaumard, B., Savard, G.: New branch-and-bound rules for linear bilevel programming. SIAM J. Sci. Stat. Comput. **13**(5), 1194–1217 (1992)
10. Vicente, L., Savard, G., Júdice, J.: Descent approaches for quadratic bilevel programming. J. Optim. Theory Appl. **81**(2), 379–399 (1994)
11. Brown, G., Carlyle, M., Diehl, D., Kline, J., Wood, K.: A two-sided optimization for theater ballistic missile defense. Op. Res. **53**(5), 745–763 (2005)
12. Wein, L.M.: Or forum-homeland security: From mathematical models to policy implementation: the 2008 philip mccord morse lecture. Op. Res. **57**(4), 801–811 (2009)
13. An, B., Ordóñez, F., Tambe, M., Shieh, E., Yang, R., Baldwin, C., DiRenzo III, J., Moretti, K., Maule, B., Meyer, G.: A deployed quantal response-based patrol planning system for the us coast guard. Interfaces **43**(5), 400–420 (2013)
14. Labbé, M., Marcotte, P., Savard, G.: A bilevel model of taxation and its application to optimal highway pricing. Manag. Sci. **44**(12), 1608–1622 (1998). part-1
15. Sinha, A., Malo, P., Frantsev, A., Deb, K.: Multi-objective stackelberg game between a regulating authority and a mining company: a case study in environmental economics. In: 2013 IEEE Congress on Evolutionary Computation (CEC), pp. 478–485. IEEE (2013)
16. Sinha, A., Malo, P., Deb, K.: Transportation policy formulation as a multi-objective bilevel optimization problem. In: 2015 IEEE Congress on Evolutionary Computation (CEC), pp. 1651–1658. IEEE (2015)
17. Nicholls, M.G.: Aluminum production modelingâĂŤa nonlinear bilevel programming approach. Op. Res. **43**(2), 208–218 (1995)
18. Hu, X., Ralph, D.: Using epecs to model bilevel games in restructured electricity markets with locational prices. Op. Res. **55**(5), 809–827 (2007)
19. Williams, N., Kannan, P., Azarm, S.: Retail channel structure impact on strategic engineering product design. Manag. Sci. **57**(5), 897–914 (2011)
20. Migdalas, A.: Bilevel programming in traffic planning: models, methods and challenge. J. Glob. Optim. **7**(4), 381–405 (1995)
21. Constantin, I., Florian, M.: Optimizing frequencies in a transit network: a nonlinear bi-level programming approach. Int. Trans. Op. Res. **2**(2), 149–164 (1995)
22. Brotcorne, L., Labbé, M., Marcotte, P., Savard, G.: A bilevel model for toll optimization on a multicommodity transportation network. Transp. Sci. **35**(4), 345–358 (2001)
23. Sun, H., Gao, Z., Wu, J.: A bi-level programming model and solution algorithm for the location of logistics distribution centers. Appl. Math. Model. **32**(4), 610–616 (2008)
24. Bard, J.F.: Coordination of a multidivisional organization through two levels of management. Omega **11**(5), 457–468 (1983)

25. Jin, Q., Feng, S.: Bi-level simulated annealing algorithm for facility location. Syst. Eng. **2**, 007 (2007)
26. Uno, T., Katagiri, H., Kato, K.: An evolutionary multi-agent based search method for stackelberg solutions of bilevel facility location problems. Int. J. Innov. Comput. Inf. Control **4**(5), 1033–1042 (2008)
27. Smith, W.R., Missen, R.W.: Chemical reaction equilibrium analysis: theory and algorithms, Wiley, xvi+ 364, 23 x 15 cm, illustrated (1982)
28. Clark, P.A., Westerberg, A.W.: Bilevel programming for steady-state chemical process designâĂŤi. fundamentals and algorithms. Comput. Chem. Eng. **14**(1), 87–97 (1990)
29. Bendsoe, M.P.: Optimization of Structural Topology, Shape, and Material, vol. 2. Springer, Berlin (1995)
30. Snorre, C., Michael, P., Wynter, L.: Stochastic bilevel programming in structural optimization (1997)
31. Mombaur, K., Truong, A., Laumond, J.-P.: From human to humanoid locomotionâĂŤan inverse optimal control approach. Auton. Robots **28**(3), 369–383 (2010)
32. Albrecht, S., Ramirez-Amaro, K., Ruiz-Ugalde, F., Weikersdorfer, D., Leibold, M., Ulbrich, M., Beetz, M.: Imitating human reaching motions using physically inspired optimization principles. In: 2011 11th IEEE-RAS International Conference on Humanoid Robots (Humanoids), pp. 602–607. IEEE (2011)
33. Bäck, T.: Evolutionary algorithms in theory and practice (1996)
34. Mathieu, R., Pittard, L., Anandalingam, G.: Genetic algorithm based approach to bi-level linear programming, Revue française d'automatique, d'informatique et de recherche opérationnelle. Recherche opérationnelle **28**(1), 1–21 (1994)
35. Yin, Y.: Genetic-algorithms-based approach for bilevel programming models. J. Transp. Eng. **126**(2), 115–120 (2000)
36. Li, X., Tian, P., Min, X.: A hierarchical particle swarm optimization for solving bilevel programming problems. Artif. Intell. Soft Comput.–ICAISC 2006, pp. 1169–1178 (2006)
37. Li, H., Wang, Y.: A hybrid genetic algorithm for solving nonlinear bilevel programming problems based on the simplex method. In: ICNC 2007 Third International Conference on Natural Computation, vol. 4, pp. 91–95. IEEE (2007)
38. Zhu, X, Yu Q., Wang, X.: A hybrid differential evolution algorithm for solving nonlinear bilevel programming with linear constraints. In: ICCI 2006 5th IEEE International Conference on Cognitive Informatics, vol. 1, pp. 126–131. IEEE (2006)
39. Sinha, A., Malo, P., Frantsev, A., Deb, K.: Finding optimal strategies in a multi-period multi-leader-follower stackelberg game using an evolutionary algorithm. Comput. Op. Res. **41**, 374–385 (2014)
40. Angelo, J.S., Krempser, E., Barbosa, H.J.: Differential evolution for bilevel programming. In: 2013 IEEE Congress on Evolutionary Computation (CEC), pp. 470–477. IEEE (2013)
41. Angelo, J.S., Barbosa, H.J.: A study on the use of heuristics to solve a bilevel programming problem. Int. Trans. Op. Res. **22**(5), 861–882 (2015)
42. Hejazi, S.R., Memariani, A., Jahanshahloo, G., Sepehri, M.M.: Linear bilevel programming solution by genetic algorithm. Comput. Op. Res. **29**(13), 1913–1925 (2002)
43. Wang, Y., Jiao, Y.-C., Li, H.: An evolutionary algorithm for solving nonlinear bilevel programming based on a new constraint-handling scheme. IEEE Trans. Syst. Man Cybern. Part C: Appl. Rev. **35**(2), 221–232 (2005)
44. Wang, Y., Li, H., Dang, C.: A new evolutionary algorithm for a class of nonlinear bilevel programming problems and its global convergence. INFORMS J. Comput. **23**(4), 618–629 (2011)
45. Jiang, Y., Li, X., Huang, C., Wu, X.: Application of particle swarm optimization based on chks smoothing function for solving nonlinear bilevel programming problem. Appl. Math. Comput. **219**(9), 4332–4339 (2013)
46. Li, H.: A genetic algorithm using a finite search space for solving nonlinear/linear fractional bilevel programming problems. Ann. Op. Res. **235**(1), 543–558 (2015)

47. Wan, Z., Wang, G., Sun, B.: A hybrid intelligent algorithm by combining particle swarm optimization with chaos searching technique for solving nonlinear bilevel programming problems. Swarm Evolut. Comput. **8**, 26–32 (2013)
48. Sinha, A., Malo, P., Deb, K.: Efficient evolutionary algorithm for single-objective bilevel optimization. arXivpreprintXiv arXiv:1303.3901 (2013)
49. Sinha, A., Malo, P., Deb, K.: An improved bilevel evolutionary algorithm based on quadratic approximations. In: 2014 IEEE Congress on Evolutionary Computation (CEC), pp. 1870–1877. IEEE (2014)
50. Angelo, J.S., Krempser, E., Barbosa, H.J.: Solving optimistic bilevel programs by iteratively approximating lower level optimal value function. In: 2013 IEEE Congress on Evolutionary Computation (CEC), pp. 470–477. IEEE (2013)
51. Bialas, W.F., Karwan, M.H.: Two-level linear programming. Manag. Sci. **30**(8), 1004–1020 (1984)
52. Chen, Y., Florian, M.: On the geometric structure of linear bilevel programs: a dual approach. Centre de Recherche sur les Transports **867** (1992)
53. Tuy, H., Migdalas, A., Värbrand, P.: A global optimization approach for the linear two-level program. J. Glob. Optim. **3**(1), 1–23 (1993)
54. Bard, J.F., Falk, J.E.: An explicit solution to the multi-level programming problem. Comput. Op. Res. **9**(1), 77–100 (1982)
55. Fortuny-Amat, J., McCarl, B.: A representation and economic interpretation of a two-level programming problem. J. Op. Rese. Soc. 783–792 (1981)
56. Ye, J.J., Zhu, D.: New necessary optimality conditions for bilevel programs by combining the mpec and value function approaches. SIAM J. Optim. **20**(4), 1885–1905 (2010)
57. Eichfelder, G.: Solving nonlinear multiobjective bilevel optimization problems with coupled upper level constraints. Inst. für Angewandte Mathematik (2007)
58. Eichfelder, G.: Multiobjective bilevel optimization. Math. Program. **123**(2), 419–449 (2010)
59. Shi, X., Xia, H.S.: Model and interactive algorithm of bi-level multi-objective decision-making with multiple interconnected decision makers. J. Multi-Criteria Decis. Anal. **10**(1), 27–34 (2001)
60. Halter, W., Mostaghim, S.: Bilevel optimization of multi-component chemical systems using particle swarm optimization. In: CEC 2006 IEEE Congress on Evolutionary Computation, pp. 1240–1247. IEEE (2006)
61. Deb, K., Sinha, A.: An efficient and accurate solution methodology for bilevel multi-objective programming problems using a hybrid evolutionary-local-search algorithm. Evol. Comput. **18**(3), 403–449 (2010)
62. Sinha, A.: Bilevel multi-objective optimization problem solving using progressively interactive emo. In: Evolutionary Multi-Criterion Optimization, pp. 269–284. Springer (2011)
63. Lai, Y.-J.: Hierarchical optimization: a satisfactory solution. Fuzzy Sets Syst. **77**(3), 321–335 (1996)
64. Sakawa, M., Nishizaki, I.: Interactive fuzzy programming for decentralized two-level linear programming problems. Fuzzy Sets Syst. **125**(3), 301–315 (2002)
65. Chen, L.-H., Chen, H.-H.: Considering decision decentralizations to solve bi-level multi-objective decision-making problems: a fuzzy approach. Appl. Math. Model. **37**(10), 6884–6898 (2013)
66. Sinha, A., Malo, P., Deb, K.: Approximated set-valued mapping approach for handling multi-objective bilevel problems. In: Working paper. IEEE (2016)
67. Deb, K., Sinha, A.: Constructing test problems for bilevel evolutionary multi-objective optimization. In: 2009. CEC'09 IEEE Congress on Evolutionary Computation, pp. 1153–1160. IEEE (2009)
68. Gupta, A., Ong, Y.-S.: An evolutionary algorithm with adaptive scalarization for multiobjective bilevel programs. In: 2015 IEEE Congress on Evolutionary Computation (CEC), pp. 1636–1642. IEEE (2015)
69. Sinha, A., Malo, P., Deb, K.: Towards understanding bilevel multi-objective optimization with deterministic lower level decisions. In: Evolutionary Multi-Criterion Optimization. pp. 426–443. Springer (2015)

70. Sinha, A., Malo, P., Deb, K., Korhonen, P., Wallenius, J.: Solving bilevel multi-criterion optimization problems with lower level decision uncertainty (2015)
71. Lu, Z., Deb, K., Sinha, A.: Handling decision variable uncertainty in bilevel optimization problems. In: 2015 IEEE Congress on Evolutionary Computation (CEC), pp. 1683–1690. IEEE (2015)
72. Deb, K., Lu, Z., Sinha. A.: Finding reliable solutions in bilevel optimization problems under uncertainties. In: 18th Annual Conference on Genetic and Evolutionary Computation, 2016. GECCO 2016. IEEE (2016)

Many-objective Optimization Using Evolutionary Algorithms: A Survey

Slim Bechikh, Maha Elarbi and Lamjed Ben Said

Abstract Multi-objective Evolutionary Algorithms (MOEAs) have proven their effectiveness and efficiency in solving complex problems with two or three objectives. However, recent studies have shown that the performance of the classical MOEAs is deteriorated when tackling problems involving a larger number of conflicting objectives. Since most individuals become non-dominated with respect to each others, the MOEAs' behavior becomes similar to a random walk in the search space. Motivated by the fact that a wide range of real world applications involves the optimization of more than three objectives, several Many-objective Evolutionary Algorithms (MaOEAs) have been proposed in the literature. In this chapter, we highlight in the introduction the difficulties encountered by MOEAs when handling Many-objective Optimization Problems (MaOPs). Moreover, a classification of the most prominent MaOEAs is provided in an attempt to review and describe the evolution of the field. In addition, a summary of the most commonly used test problems, statistical tests, and performance indicators is presented. Finally, we outline some possible future research directions in this research area.

Keywords Many-objective optimization · Evolutionary algorithms · Scalability · High dimensionality

S. Bechikh (✉) · M. Elarbi · L. Ben Said
SOIE Lab, Computer Science Department, ISG-Tunis, University of Tunis,
Bouchoucha city, 2000 Le Bardo, Tunis, Tunisia
e-mail: slim.bechikh@gmail.com

M. Elarbi
e-mail: arbi.maha@yahoo.com

L. Ben Said
e-mail: lamjed.bensaid@isg.rnu.tn

S. Bechikh et al. (eds.), *Recent Advances in Evolutionary Multi-objective Optimization*, Adaptation, Learning, and Optimization 20,
DOI 10.1007/978-3-319-42978-6_4

1 Introduction

Since the implementation of the first MOEA, different algorithms have been proposed to deal with Multi-objective Optimization Problems (MOPs) [1]. MOEAs have been widely used to solve problems with two or three objectives. In fact, most of the proposed MOEAs use the Pareto-dominance relation to compare solutions of the population. Specially, the population members are ranked using the Pareto-dominance relation and the recombination operator is performed to the best individuals in order to generate solutions that are closer to the Pareto set. However, recent studies on MOEAs have shown that Pareto-based MOEAs struggle to solve problems with more than three objectives. Thus, although the classical MOEAs such as NSGA-II [2] and SPEA2 [3] have been successfully applied to solve many real-world problems with a small number of objectives, they are not well-suited when dealing with problems involving a high number of objectives. This limitation seems to affect only Pareto-based MOEAs but some difficulties are common to most existing multi-objective optimizer. For this reason, motivated by the fact that a wide range of real world applications in industrial [4] and engineering [5] designs involves the optimization of more than three objectives, a wide variety of proposals have been proposed to deal with the difficulties encountered by the current state of the art MOEAs.

In summary, the challenges encountered by the state of the art MOEAs in finding a representative set of Pareto optimal solutions when handling MaOPs can be briefly discussed as follows:

- **Increase of the number of non-dominated solutions**: When the dimensionality of the objective space increases, the proportion of Pareto-non dominated solutions in the population grows which deteriorates the search process ability to converge towards the Pareto front. Thus, the MOEA behavior becomes similar to a random search one. Figure 1 shows how the proportion of non-dominated solutions in the population behaves with respect to the number of objectives. We can see that after a few generations, the population becomes completely non-dominated.

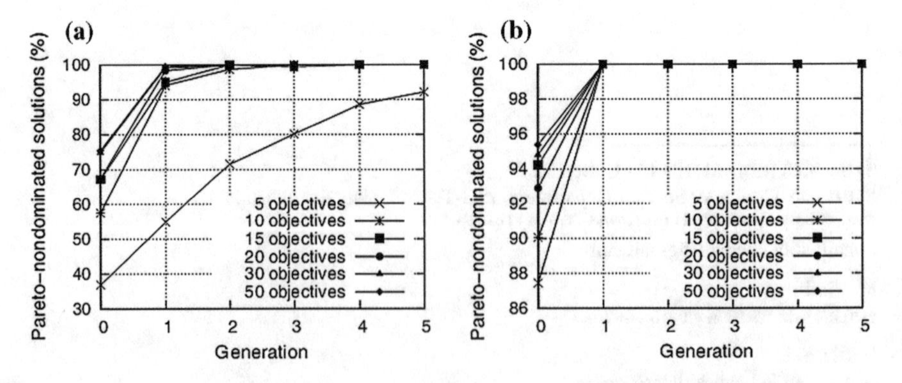

Fig. 1 Proportion of Pareto-non-dominated solutions. From Ref. [6]. **a** DTLZ1. **b** DTLZ6

Table 1 Bounds for the number of points required to represent a Pareto front with resolution $r = 25$

M	Points
2	50
4	62 500
5	1 953 125
7	1 708 984 375

From Ref. [7]

As a consequence, new promising search directions become very hard to find. Another reason is the increment of the number of dominance resistant solutions in the population when we deal with many-objective. In fact, dominance resistant solutions are non-dominated solutions but they are far from the True Pareto Front.

- **Ineffectiveness of crossover and mutation operators**: In a high dimensional space, the population members are likely to be widely distant from each other. Thereby, two distant parent solutions will produce two distant children that are not similar to their parents. In such a case, the effect of the recombination operation becomes inefficient in producing promising offspring individuals.
- **Difficulty to represent the trade-off surface**: Due to the high dimensionality, more points are needed to represent the trade-off surface. In fact, the number of points to represent a Pareto front with M objectives and r resolution is bounded by $O(Mr^{M-1})$. This expression is derived assuming that each solution is contained in a hypercube. Thus, the resolution r represents the number of hypercubes per dimension. Table 1 shows the bound of points required to represent a Pareto front for different number of objectives using a resolution $r = 25$. We note that for 5 objectives the number of points required to represent the Pareto front is about 2 million points.
- **High computational cost of the diversity measure estimation**: In order to determine the extent of crowding of a solution in a population, the identification of neighboring solutions in a population becomes computationally very expensive in high dimensional spaces. For this reason, the use of any approximation in diversity to reduce the computational cost may cause an unacceptable distribution of the solutions.
- **Difficulty of visualization**: It is not a matter that is directly related to optimization. The visualization of a higher dimensional trade-off front becomes difficult. Hence, it is difficult for the Decision Maker (DM) to choose a preferred solution. Several methods were proposed to ease decision making in MaOPs such as Parallel coordinates and self-organizing maps.

2 A Taxonomy of Many-objective Solution Approaches

In this section, a classification of the most relevant approaches to deal with MaOPs is presented.

2.1 Preference Ordering Relation-Based Approaches

- **Expansion Relation**

The Expansion preference relation (ER) was proposed by Sato et al. [8] to control the dominance area of solutions using a user-defined parameter S. This preference ordering relation was proposed in order to induce an appropriate ranking of solutions and to enhance the selection mechanism, so that the performance of MOEAs on combinatorial optimization problems with a variety of objectives is improved. The basic idea consists of expanding and contracting the dominance area by replacing the objective function $f_i(x)$ using the vector S as follows:

$$f_i^{'}(x) = \frac{r\ sin(\omega_i + S_i\ \Pi)}{sin\ (S_i\ \Pi)} \qquad \forall i \in \{1, 2, \ldots, m\} \tag{1}$$

where r is the norm of $f(x)$, $f_i(x)$ is the fitness value of the i-th objective, and ω_i is the angle between $f(x)$ and $f_i(x)$. Figure 2 illustrates the fitness modification to change the covered area of dominance when $S_i < 0.5$ and $\phi_i = S_i\ \Pi$. One can notice that the i-th fitness value $f_i(x)$ is increased to $f_i^{'}(x) > f_i(x)$. Thus, if $S_i < 0.5$ a more finer grained ranking of solutions is produced and the dominance area is expanded which strengthen the selection. However, if $S_i > 0.5$ a coarser ranking of solutions is produced and the dominance area is contracted which would weaken the selection. While if $S_i = 0.5$, the usual dominance relation is used. Since in a MaOP we search to produce a finer grained ranking of solutions, the parameter S_i should be less than 0.5. In fact, the main characteristic of this preference relation is that it emphasizes the solutions in the middle region of the Pareto front. The authors used the multi-objective 0/1 Knapsack problem [9] on two up to five objectives and integrated the ER relation into NSGA-II. The experimental results show that contracting or expending the dominance area is better than using conventional dominance in terms of the quality of the obtained solutions. However, the ER was assessed only on problems involving up to five objectives. Hence further experiments with higher dimension problems are

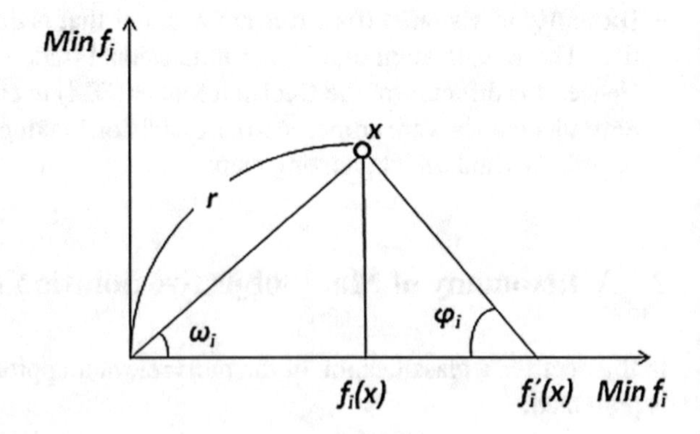

Fig. 2 Fitness modification to change the covered area of dominance for $S_i < 0.5$. From Ref. [8]

required for validation. Moreover, since the expansion relation can improve either convergence or diversity, the authors concluded that it would be better to combine it with other methods.

- ***k*-Optimality Relation**

Farina and Amato [10] proposed the k-optimality relation. This preference relation is based on the number of improved objectives between two solutions. The k-optimality employs three quantities. Assuming that we have two solutions x and y, the first quantity n_b represents the number of objectives where x is better than y. The second one n_e denotes the number of objectives where x is equal to y and the final one n_w where x is worse. Thus, given M objectives the following inequalities holds true:

$$n_b + n_w + n_e = M \tag{2}$$

$$0 < n_b, n_w, n_e < M \tag{3}$$

In fact, by employing these quantities the concepts of (1-k)-dominance and k-optimality can be defined.

Definition 1 *(1-k)-dominance*
A solution x (1-k)-dominates a solution y if and only if:

$$\begin{cases} n_e < M \\ n_b \geq \frac{M-n_e}{1+k} \end{cases} \quad 0 \leq k \leq 1 \tag{4}$$

From the above definition, one can notice that the 1-dominance (i.e., $k = 0$) represents the Pareto dominance. The parameter k can assume any value in [0,1], but because n_b has to be a natural number, the smallest integer greater than the quantity $\frac{M-n_e}{1+k}$ need to be considered. After defining the (1-k)-dominance, the following definition of the k-optimality can be given:

Definition 2 *k-optimality*
A solution x^* is k-optimum if and only if there is no $x \in \Omega$ such that x k-dominates x^*.

Therefore, the k-optimality represents a strong version of the Pareto-optimality (0-optimality). The authors extended the (1-k) dominance relation by incorporating fuzzy arithmetic techniques.

- **Average and Maximum Ranking Relations**

Bentley and Wakefield [11] proposed the average ranking (AR) and maximum ranking (MR) preference ordering relations. The AR relation begins by sorting the solutions based on their fitness. Then, a set of different ranking for every solution is obtained for each objective. After that, the average ranking value of each solution is computed by summing their ranks on each objective. Hence, based on the obtained average ranking values, the solutions can be sorted into order of best average rank. Thus, a solution x dominates a solution y with respect to the AR relation (denoted by $x \prec_{avg} y$) if and only if $R_{avg}(x) < R_{avg}(y)$ where $R_{avg}(x) = \sum_{1 \leq i \leq M} f_i(x)$. The AR distinguishes the non-dominated solutions based on their obtained ranks on different objectives. This preference ordering relation is simple and range-independent [12]. However, Corne and Knowles [13] have reported that the obtained solution set may only concentrate in a subregion of the Pareto front. Thus, it has a lack of diversity maintenance mechanism.

Differently, the MR relation considers the best rank as the global rank of each solution. Therefore, a solution x dominates a solution y with respect to the MR relation (denoted by $x \prec_{max} y$) if and only if $R_{max}(x) < R_{max}(y)$ where $R_{max}(x) = min_{1 \leq i \leq M} \{rankf_i(x)\}$. The main drawback of this method is that it emphasizes solutions with high performance in some objectives, while they have a poor overall performance (i.e., extreme solutions).

- **Favour Relation**

In order to refine the ranking of solutions in MaOPs, Drechsler et al. [14] proposed the favour relation. In their work, the authors used the favour relation and a method called Satisfiability Class Ordering (SCO) where the former is used to compare solutions to each others, while the latter is used to sort solutions. This preference ordering relation can be defined as follows:

Definition 3 *Favour relation*
A solution x dominates a solution y with respect to the favour relation (denoted by $x \prec_f y$) if and only if:

$$|\{i : f_i(x) < f_i(y), 1 \leq i \leq M\}| < \left|\left\{j : f_j(y) < f_j(x), 1 \leq j \leq M\right\}\right| \quad (5)$$

The main idea behind the favour relation is that the solution x is favoured to y if and only if the number of objectives in which x outperforms y is superior to the number of objectives in which y outperforms x. For example given two solutions x_1 $(4, 2, 1)$, and x_2 $(1, 1, 2)$ then we have that $x_2 \prec_f x_1$ since it has better objective values than x_1 with respect to the first two objectives. This preference ordering relation was used in an algorithm proposed by Drechsler et al. [15]. In fact, it was demonstrated that the favour relation does not create a partial order since it is not transitive but it is able to create a finer grained ranking of solutions than that created by the Pareto dominance when solving MaOPs. However, the main disadvantage of this relation is that it emphasizes extreme solutions.

- **Other Preference Ordering Relation-Based Approaches**

Preference ordering relation-based approaches have been proposed to deal with the first challenge which is the increase of the number of Pareto non-dominated solutions in the population in a high dimensional space. Therefore, preference ordering relation-based approaches aim mainly to provide a finer ranking of solutions when solving MaOPs. Various methods based on preference ordering relations have been proposed to deal with MaOPs. The preference order ranking was introduced by Di Pierro et al. [16]. The basic idea of this preference relation is to discard the objectives in order to compare the solutions. It is based on the concept of *efficiency of order* proposed by Das [17]. However, the main drawback of this relation is its high computational cost. Sülflow et al. [4] proposed the ε-preferred relation which is based on the favour relation [14]. In the ε-preferred relation, two solutions are compared based on the number of objectives in which one solution exceeds the other using a predefined threshold. Moreover, the favour relation is used to determine which solution is better in case of a tie. The authors replaced the favour relation in the algorithmic framework used in [14] with the ε-preferred relation. The ε-preferred relation has demonstrated good results on the nurse rostering problem [18] with 25 objectives. A summary of some existing preference ordering relations for solving MaOPs is presented in Table 2.

Table 2 Summary of preference ordering relations: MaxObjs means maximum number of objectives

Relations	References	Basic idea	MaxObj	Test problems
ER	Sato et al. [8]	Control the dominance area of solutions using a user defined parameter *S*	5	MKP
k-Optimality	Farina and Amato [10]	Compare two solutions based on the number of improved objectives between them	12	Test case
AR	Garza-Fabre et al. [6]	Sort the solutions based on their average ranking values	50	DTLZ
MR	Garza-Fabre et al. [6]	Compare the solutions based on their best obtained rank	50	DTLZ
Favour	Drechsler et al. [14]	Favour the solution that outperforms the other one in more objectives	7	5 benchmarks problems
Preference order ranking	Di Pierro et al. [16]	Compare two solutions by discarding objectives	8	DTLZ
ε-Preferred	Sülflow et al. [4]	Compare two solutions based on the number of objectives in which a solution exceeds the other one by using a predefined threshold	25	NRP

2.2 *Objective Reduction-Based Approaches*

- **PCA-NSGA-II: Principal Component Analysis-NSGA-II**

In this work, Deb and Saxena [19] proposed the Principal Component Analysis-Nsga-II algorithm called PCA-NSGA-II. This latter combines a reduction technique with NSGA-II to deal with MaOPs with redundant objectives. In fact, many real world problems have M objectives, while the true Pareto front is less than M-dimensional. Hence, some of the objectives are redundant. Thus, in order to determine the true Pareto optimal front, the authors suggested to use the PCA procedure. This reduction technique was used to reduce the dimensionality of a data set with a large number of interrelated variables. The PCA-NSGA-II starts with an initial set of objectives $\Pi_0 = \{1, 2, \ldots, M\}$. Then, NSGA-II is executed for a given number of iterations to obtain a population P_t where t represents the current generation. Next, the population P_t is used by the PCA reduction method to get a new set of objectives $\Pi_t \subset \Pi_0$ to be used in the next iterations of NSGA-II. The PCA procedure can be summarized by the following four basic steps:

- **Step 1**: Store the objective values of the population P_t in an initial data matrix D of size $M \times N$, where M is the number of objectives and N is the size of the population;
- **Step 2**: Obtain the standardized matrix X by subtracting the mean from each objective value in matrix D;
- **Step 3**: Compute the covariance matrix V and the correlation matrix R using the standardized matrix X;
- **Step 4**: Compute eigenvalues of the correlation matrix R and eigenvectors that are considered as the PCs;

We note that the most negative and the most positive elements for a given PC are considered as the two most important conflicting objectives. In addition, PCA-NSGA-II approach uses an additional procedure that selects the most negative and most positive elements for the first PC. The experimental results on a modified version of DTLZ5 test problem [19] have demonstrated the ability of PCA-NSGA-II to solve high dimensional problems with redundant objectives. However, the proposed algorithm shows some vulnerability when the task involves finding a large Pareto optimal front due to the difficulties encountered to find the correct combination of objectives in MaOPs with non-redundant objectives.

- **PCSEA: Pareto Corner Search Evolutionary Algorithm**

Singh et al. [20] introduced the Pareto Corner Search Evolutionary Algorithm (PCSEA). The authors proposed a new approach that identifies a reduced set of objectives instead of dealing with the true dimensionality of the true MaOP. Moreover, PCSEA does not approximate the whole Pareto front but it searched for a specific set of non-dominated solutions. More specifically, the authors suggested to use boundaries of the Pareto front called corner solutions in order to predict the dimensionality of the true Pareto front. In fact, for a two-dimensional optimization problem, a corner solution corresponds to the minimum value of each objective. However, the number of corner solutions increases exponentially with the number of objectives (i.e., $2^M - 1$ possible corners to a M-objective optimization problem), but in reality, test problems such as DTLZ [21] and WFG [22] have M corner solutions for M-objective problems. The proposed approach can be summarized by the following two steps: (1) find the corner solutions, and (2) use the corner solutions to reduce the set of objectives. PCSEA uses the same crossover and mutation operators used in NSGA-II. However, differently to NSGA-II which uses non-dominated sorting and crowding distance-based ranking, PCSEA uses a corner-sort ranking. Details of this method can be found in [20]. After identifying the corner solutions, a heuristic technique is performed to determine the relevant objectives and to eliminate the redundant ones. The reduction process can be described as follows. First, a set F containing the non-dominated solutions produced by PCSEA is formed where only unique solutions are considered in the set F. Second, in order to quantify the change in the number of non-dominated solutions a parameter R was used. The parameter R is defined as follows:

$$R = N_{F_R - f_m} / N_F \tag{6}$$

where N_F is the number of non-dominated solutions in the set F and $N_{F_R - f_m}$ is the number of non-dominated solutions corresponding to the objective set obtained after omitting f_m from the set of relevant objectives F_R. If the value of R is high for a particular objective f_m, it means that this objective can be omitted from the set of relevant objectives. PCSEA does not suffer from the lack of the selection pressure and it has a low computational complexity which makes it suitable for solving MaOPs. However, it should be noted that a large population size is not required when dealing with many objectives for the reason that PCSEA does not approximate the entire Pareto front.

- **Objective Reduction Using a Feature Selection Technique**

In this work, López Jaimes et al. [23] proposed to integrate an unsupervised feature selection technique that was originally introduced by Mitra et al. [24] in NSGA-II. This reduction method is similar to the one used by [19] for the reason that both of them use a correlation matrix to measure the conflict between each pair of objectives and to determine the most conflicting objectives in order to eliminate the redundant ones. Two algorithms have been introduced in this work. The first algorithm finds the minimum subset of non-redundant objectives with the minimum possible error, while the second algorithm finds the minimum set of k-non-redundant objectives that yield to the minimum possible error. The authors described the main steps of their reduction technique by the following three steps:

- **Step 1**: Define the conflict between objectives as distance and divide the objective set into homogeneous neighborhoods of size q around each objective;

- **Step 2**: Select the most compact neighborhood where the most compact neighbors is the neighborhood with the minimum distance to its qth neighbor;
- **Step 3**: Retain the center of the neighborhood and discard q neighbors with least conflict in the current set. The distance to the qth neighbor is considered as the error committed by removing the q objectives;

The reduction techniques used in this work iterated **Step 2** and **Step 3** until the number of desired objectives does not reach the predefined k value or until there are not more considered neighborhoods. The experimental study has mentioned good results in solving: a variation of DTLZ5 [21] with 3, 5, and 10 objectives, a variation of the DTLZ2 [21], and the 0/1 knapsack problem [9] with 10 and 20 objectives. The experiment results show that the proposed methods are competitive compared to the PCA-based reduction method and the reduction method of Brockhoff et al. [25].

- **Other Objective Reduction-Based Approaches**

Objective reduction-based approaches aim to tackle MaOPs with redundant objectives. In fact, there are two different timing of incorporating the dimensionality into a MOEA, thus we can identify two classes [28]: (1) offline dimensionality reduction where the dimensionality reduction method is performed after obtaining a set of Pareto optimal solutions or (2) online dimensionality reduction where the number of objectives is introduced gradually by iteratively obtaining solution sets and invoking the dimensionality reduction method during the search process. In fact, the first class can be further divided into three sub-classes: Correlation-based methods, dominance structure-based methods, and feature-based methods. Correlation-based methods consist in examining the correlation among the objectives. In this sub-class, we find the work of Saxena et al. [26] in which they proposed L-PCA and NL-MVU-PCA algorithms based on the PCA method and Maximum Variance Unfolding for linear and nonlinear objective reduction, respectively. The authors investigated the performance of the two algorithms on a wide range of redundant and non-redundant test problems and on two real world problems. Dominance structure-based methods consider the dominance relationships among the solutions obtained by a MOEA. Brockhoff and Zitzler [27] proposed a new notion of conflict and they introduced a quantification δ for measuring the change in the dominance structure based on ε-dominance. In their studies, an exact and a greedy algorithm were proposed to solve the δ-MOSS and the k-EMOSS problems, where the δ-MOSS consists in finding the minimum objective subset corresponding to a given error, while k-EMOSS consists in finding an objective subset of size k with the minimum possible error. The experimental results demonstrated that the exact algorithm yields smaller objective subsets than the greedy algorithm, while the high complexity of the exact method limits its usage. In the third sub-class, we find the work of López Jaimes et al. [23]. Concerning the online dimensionality reduction algorithms, many approaches have been proposed in the literature. PCA-NSGA-II is considered as an online dimensional reduction algorithm since it iteratively obtains solution sets and reduces the objectives using information of correlations among the objectives. Table 3 presents a comparison of the objective reduction-based approaches studied in this subsection.

Table 3 Comparison of objective reduction-based approaches: OnDRA means online dimensionality reduction approach, OfDRA means offline dimensionality reduction approach, C means correlation-based methods, DS means dominance structure-based methods, F means feature-based methods, and MObj mcans maximum number of objectives

Algorithms	References	Characteristics					MObj	Test problems
		OnDRA	OfDRA					
			C	DS	F	Other		
PCA-NSGA-II	Deb and Saxena [19]	X	–	–	–	–	30	DTLZ5(I,M)
							5	DTLZ2
PCSEA	Singh et al. [20]	–	–	–	–	X	100	DTLZ5(I,M)
							20	DTLZ2
–	López Jaimes et al. [23]	–	–	–	X	–	20	MKP
							20	$DTLZ2_{BZ}$
							10	DTLZ5(I,M)
L-PCA	Saxena et al. [26]	–	X	–	–	–	25	DTLZ
							25	WFG
							50	DTLZ5(I,M)
NL-MVU-PCA	Saxena et al. [26]	–	X	–	–	–	25	DTLZ
							25	WFG
							50	DTLZ5(I,M)
Exact-δ-MOSS/k-EMOSS	Brockhoff and Zitzler [27]	–	–	X	–	–	–	–
Greedy-δ-MOSS/k-EMOSS	Brockhoff and Zitzler [27]	–	–	X	–	–	25	DTLZ
							25	MKP

2.3 *Preference Incorporation-Based Approaches*

- **R-NSGA-II: Reference Point-Based NSGA-II**

Deb et al. [29] introduced a modified version of NSGA-II that prefers solutions closer to a user-provided reference point set and that de-emphasizes solutions within a ε-neighborhood of a reference point. In fact, the parameter ε controls the extent of the distribution of solutions near the closest Pareto-optimal solution and the reference points are used to guide the search toward the preferred parts of the Pareto front. In this work, the crowding distance used in NSGA-II is modified as follows. For each reference point, the normalized Euclidean distance of each solution of the last considered front is calculated and based on this distance the solutions are sorted in ascending order. Hence, the closest solution to the reference point is assigned a rank of one. The second closest solution to the reference point is assigned a rank of two and so on. After that, the minimum of the assigned ranks is assigned as the crowding distance to a solution. Thus, the smallest crowding distance of one is assigned to the closest solutions to all reference points. The solutions having next-to-smallest Euclidean distance to all reference points are assigned the next-to-smallest crowding distance of two, and so on. Thereafter, solutions with a smaller crowding distance are preferred. The authors proposed to control the extent of obtained solutions by grouping all solutions having a sum of normalized difference in objective values of ε or less. A randomly picked solution from each group is retained and the rest of all group members are assigned a large crowding distance in order to discourage them to remain in the race. The proposed procedure allows finding multiple ROIs simultaneously in a single simulation run. R-NSGA-II has demonstrated good results on two to five objective test problems but it faces difficulties when using a single reference point since diversity is not well maintained. Moreover, the ε clearing parameter setting is not trivial.

- **PBEA: Preference-Based Evolutionary Algorithm**

In this work, Thiele et al. [30] proposed a new algorithm called PBEA that combines IBEA with the reference point method. In fact, in IBEA, the fitness value of a solution x in a population P can be expressed as follows:

$$F(x) = \sum_{y \in P\{x\}} (-\mathrm{e}^{-I(y,x)}/\kappa) \tag{7}$$

where κ is a scaling factor. In IBEA, the additive epsilon indicator is used which is a Pareto compliant indicator and it is defined as follows:

$$I_{\varepsilon^+}(x, y) = min_{\varepsilon}\{f_m(x) - \varepsilon \leq f_m(y)\ \forall m = 1, 2, \ldots, M\} \tag{8}$$

In order to take the preference information into account, the authors defined a new preference-based quality indicator described as follows:

$$I_p(x, y) = I_{\varepsilon^+}(x, y)/s(g, f(x), \delta) \tag{9}$$

where x and y are two solutions, I_{ε^+} is the additive epsilon indicator, s is a function used to normalize the set of points, and δ is a positive parameter used to specify the minimal value of the normalized function, it allows the DM to control the spread of the obtained Region of Interest (ROI). PBEA was also used in an interactive fashion to offer many possibilities to the DM in directing the search into a preferred part of the Pareto optimal set. The main motivation behind PBEA is that it gives reliable information on the solutions to the DM. Moreover, the used binary quality indicator I_p is Pareto dominance preserving. In addition, the experimental results show that it is suitable to solve MaOPs due to the use of the achievement function. However, the authors noted that adjusting the δ parameter is not an easy task.

- **r-NSGA-II: Reference Solution-Based NSGA-II**

Ben Said et al. [31] proposed a new dominance relation called r-dominance (reference-solution-based dominance) that creates a strict partial order among Pareto equivalent solutions and that has the ability to differentiate between non-dominated solutions in a partial manner based on a user-supplied aspiration level vector. The r-dominance relation represents a hybridization between the Pareto dominance relation and the reference point method (i.e., DM' s preferences). In fact, the key feature of this preference-based dominance relation is to prefer solutions that are closer to the reference point, while preserving the order induced by the Pareto dominance. Thus, in order to determine the closeness of a solution to the reference point, the authors used the weighted Euclidean distance employed by Deb et al. [30] which is expressed as follows:

$$dist(x, g) = \sqrt{\sum_{i=1}^{M} w_i \left(\frac{(f_i(x) - f_i(g))}{(f_i^{max} - f_i^{min})} \right)^2} \quad w_i \in]0, 1[\ \sum_{i=1}^{M} w_i = 1 \tag{10}$$

where x is a solution, g is a reference point, f_i^{max} and f_i^{min} represent the upper and the lower bounds of the i-th objective, respectively, and w_i is the weight associated to each objective. The r-dominance is defined as follows:

Definition 4 *r-dominance*
Assuming a population of individuals P, a reference vector g, and a weight vector w, a solution x is said to r-dominate a solution y (denoted by $x \prec_r y$) if one of the following statements holds true:

1. x dominates y in the Pareto sense,
2. x and y are Pareto-equivalent and $D(x, y, g) < -\delta$, where $\delta \in [0, 1]$ and:

$$d(x, y, g) = \frac{dist(x, g) - dist(y, g)}{dist_{max} - dist_{min}} \tag{11}$$

$$dist_{max} = Max_{z \in P}\ dist(z, g) \tag{12}$$

$$dist_{min} = Min_{z \in P}\ dist(z, g) \tag{13}$$

δ is termed as the non-r-dominance threshold.

After substituting the Pareto dominance with the r-dominance in the NSGA-II algorithm with an adaptive management of the δ parameter, the performance of the resulting preference-based MOEA, named r-NSGA-II, has been assessed on several test problems where the number of objectives is varying between two and ten objectives. The experimental results show that r-NSGA-II outperforms several recent reference point approaches. Moreover, the r-dominance was able to guide the search using the DM' s preferences and to control the spread of the region of interest. However, r-NSGA-II algorithm has faced difficulties in solving highly multi-modal problems such as ZDT4 [32].

- **PICEA-g: Preference-Inspired Co-Evolutionary Algorithm-Goals**

PICEA-g was introduced by Wang et al. [33]. PICEA-g is a posteriori preference-based algorithm where the intervention of the DMs is performed after obtaining a solution set which approximates the real Pareto front. The main idea of this algorithm is to provide DMs with both a proximal and a diverse representation of the entire Pareto front before the elicitation and the application of their preferences. As the search progress, PICEA-g coevolves a family of DM' preferences together with a population of candidate solutions. Thus, the solutions would gain fitness by performing well against the preferences and the preferences would gain fitness by offering comparability between solutions. The general principle of PICEA-g is as follows. The PICEA-g begins by initializing a population of candidate solutions S and preference sets G of fixed size N and $NGoal$, respectively. In each generation t, genetic variation operators are applied to the parents $S(t)$ in order to produce N offspring $Sc(t)$. Simultaneously, $NGoal$ new preference sets $Gc(t)$, are randomly regenerated based on the initial bounds. Thereafter, $S(t)$ and $Sc(t)$ and both $G(t)$ and $Gc(t)$ are then pooled, respectively. After that, the obtained populations are sorted based on the fitness. Finally, a truncation selection is applied to select N solutions to form the new parent population $S(t+1)$ and $NGoal$ solutions as new preference population $G(t + 1)$. The method to calculate the fitness F_s, of a candidate solution s is defined as follows:

$$F_s = 0 + \sum_{g \in G \uplus G_c | s \preccurlyeq g} \frac{1}{n_g} \tag{14}$$

where n_g is the number of solutions that satisfy preference g. It should be noted that if s does not satisfy any g, then F_s is equal to zero. The fitness F_g of a preference g can be expressed as follows:

$$F_g = \frac{1}{1+\alpha} \tag{15}$$

where

$$\alpha = \begin{cases} 1 & \text{if } n_g = 0 \\ \frac{n_g - 1}{2N-1} & \text{otherwise} \end{cases} \tag{16}$$

where N is the candidate solution population size. After calculating fitness values, the non-dominated solutions in $S \cup S_c$ are identified. Then, based on the fitness, the best N non-dominated solutions are selected to constitute the new parent $S(t + 1)$. The authors reported that PICEA-g outperforms several state-of-the-art methods in terms of convergence and spread when compared on WFG test problems with up to 10 objectives.

- **Other Preference Incorporation-Based Approaches**

In the context of incorporating preference information in EMO, many studies have been made [34, 35]. A key point in preference-based approaches is the timing of integrating the preference information into the optimizing process. In fact, the DM can provide his/her preferences before (a priori), after (a posteriori), or during (interactively) the MOEA run [36–38]. Since the search direction is biased towards the area of the Pareto front on which the DM would like to focus (i.e., ROI), the priori and interactive algorithms can reduce the computational load during the search. However, the posteriori preference-based algorithms are inferior to the above mentioned classes since they might obtain a large number of solutions that the DM is not interested in [28]. Deb and Kumar [40] proposed the reference direction-based NSGA-II (RD-NSGA-II). In each iteration, the DM supplies a reference direction in the objective space. Thereafter, the solutions are ranked using an achievement scalarizing function and the crowding distance value. RD-NSGA-II has demonstrated good results when tested on DTLZ functions with up to 10 objectives. However, the population diversity degradation that can be yielded when using a single reference direction remains a significant matter. Preference-inspired co-evolutionary algorithms (PICEAs) represent an example of a posteriori preference-based algorithm that tries to avoid the intervention of the DM before or during the optimization process. In PICEAs preferences are modeled as a set of solutions which co-evolve with the population [41, 42]. In [41, 42], the authors tested a-PICEA-g and PICEA-w on the WFG test problems with up to 7 objectives. Table 4 presents a comparison of the studied preference incorporation-based approaches that are classified into three main classes: (1) Priori preference-based approaches, (2) Interactive preference-based approaches, and (3) Posteriori preference-based approaches.

2.4 *Indicator-Based Approaches*

- **IBEA: Indicator-Based Evolutionary Algorithm**

IBEA was introduced by Zitzler and Künzli [43]. They proposed a general IBEA where they used a binary performance indicator in the selection process. Initially, IBEA begins by generating an initial population P. Then, the algorithm calculates the fitness value of each solution x in P. In fact, the fitness value is a measure for the loss in quality if a solution x is removed from P. After computing all the fitness values of all individuals in the population, the worst individual is removed from the population and the fitness values of the residual population must be updated. In the following, the selection step is used in creating the mating pool P'. When we compare IBEA with the use of two binary performance indicators the additive ε-indicator and the I_{HD}-indicator to Pareto-based MOEAs such as SPEA2 and NSGA-II, we note that IBEA can greatly improve the quality of the generated Pareto set approximation. In addition, IBEA outperforms NSGA-II and SPEA2 in term of convergence. However, the parameter κ which is a scaling factor of the fitness function values should be

Table 4 Comparison of preference incorporation-based approaches for MaOPs: MObjs means maximum number of objectives (inspired by [28, 31])

Classes	Algorithms	References	MObjs	Test problems	Preference information
Priori preference-based approaches	PBEA	Thiele et al. [30]	5	LPMS	Reference points
	SBGA	Gong et al. [39]	20	DTLZ	Preferred regions
Interactive preference-based approaches	R-NSGA-II	Deb et al. [29]	10	DTLZ	Reference points
	r-NSGA-II	Ben Said et al. [31]	10	DTLZ	Reference points
	RD-NSGA-II	Deb and Kumar [40]	10	DTLZ	Reference directions
Posteriori preference-based approaches	PICEA-g	Wang et al. [33]	10	WFG	Weight vectors
	a-PICEA-g	Wang et al. [41]	7	WFG	Goal vectors
	PICEA-w	Wang et al. [42]	7	WFG	Weight vectors

appropriately chosen. The main weakness of IBEA is the computational cost of the quality indicator value. Several variants of IBEAs have been proposed such as the work of Basseur and Bruke [44] in which they extended IBEA and proposed a multi-objective local search algorithm called IBMOLS that uses a local search operator and the work of Wagner et al. [45] that reported good results for MaOPs. Since, IBEAs do not use Pareto dominance, their search ability is not severely deteriorated by the increase of the number of objectives. It should be noted that most of the existing variants use the hypervolume as an indicator but one difficulty arises in using the hypervolume when dealing with a large number of objectives which is the high computational cost of the hypervolume calculation.

- **SMS-EMOA: *S* Metric Selection-Based Evolutionary Multi-objective Algorithm**

One of the most successfully used indicator-based MOEAs, is the S-Metric-Selection-EMOA (SMS-EMOA) proposed by Emmerich et al. [46]. The SMS-EMOA invokes firstly the non-dominated sorting that is used as a ranking criterion. Secondly, it uses the hypervolume indicator as a selection mechanism to discard the individual that contributes the least hypervolume to the worst-ranked front. The SMS-EMOA algorithm starts with generating a new population P with μ individuals. In each iteration, there is a new individual that is generated by the application of random variation operators. An individual becomes a member of the population if it replaces dominated individuals and contributes to a higher quality of the population. Thus, the selection criterion ensures that the non-dominated individuals could not be replaced by the dominated ones. Then, the algorithm applied the fast-non-dominated-sort-algorithm used in NSGA-II to compute the Pareto fronts. After that, an individual is rejected from the worst ranked front R_I if it contains more than one individual. Thus, the individual $n \in R_I$ that minimizes the following equation is discarded:

$$\Delta_S(n, R_I) = S(R_I) - S(R_I\{n) \quad (17)$$

where the $\Delta_S(n, R_I)$ represents the contribution of n to the S metric value of its appropriate front. The application of this algorithm to several standards benchmark shows that it is suitable for Pareto optimization with two and three objectives. Rather than that, SMS-EMOA outperforms a number of Pareto-based algorithms in term of convergence. It is also shown that it provides solutions that are well distributed on the Pareto Front. The main disadvantage of this indicator based-MOEA is the high computational coast of the S-metric values with problems evolving more than three objectives. Moreover, SMS-EMOA is well-suited for real-world applications with a limited number of function evaluations. Wagner and Neumann [47] have compared SMS-EMOA to a number of Pareto-based algorithms and indicator-based algorithms on MaOPs. The results show that SMS-EMOA is unable to find the front of the high-dimensional DTLZ1 and DTLZ3 test problems.

- **AGE: Approximation-Guided Evolutionary**

AGE was proposed by Bringmann et al. [48]. AGE uses the additive approximation. In fact, the additive approximation of the set B with respect to the set A is expressed as follows:

$$\alpha(A, B) = \max_{a \in A} \min_{b \in B} \max_{1 \leq i \leq N} (a_i - b_i) \quad (18)$$

It could also use the multiplicative approximation which is similar to the additive approximation by just replacing $a_i - b_i$ with $\frac{a_i}{b_i}$. The goal is to minimize the additive approximation that measures the approximation quality of the population B with respect to the archive A. The archive A contains all non-dominated solutions seen so far. However, the additive approximation is not locally sensitive to the changes of the output population. AGE uses another sensitive indicator that should be minimized which is defined as follows:

$$S_\alpha(A, B) = (\alpha_1, \ldots, \alpha_{|A|}) \quad (19)$$

where $S_\alpha(A, B)$ is the result of sorting decreasingly the set $\alpha(\{a\}, B)|a \in A$. The algorithm begins by generating a population P of μ individuals. In each iteration, we obtain λ new offspring by selecting randomly two individuals from the population and applying the crossover and the mutation operators. Those λ new offspring individuals are added to P and a new population Q is obtained. After that, only non dominated solutions obtained from Q are added to A. In addition, there are two criteria to add a solution S to A: (1) S is not dominated by any existing individual in A and (2) individuals that dominate S are removed. In each generation, the individual p with lexicographically worst approximation is removed from Q. AGE was compared to several MOEAs and as a result it was proved that AGE outperforms them in term of the quality of the approximation set obtained especially when dealing with many objectives and the covered hypervolume. Wagner and Neumann [47] extended AGE and presented a new version called AGE-II where they control the size of the archive by storing the additive ε-approximation of the non-dominated solutions and they propose a new strategy for the parent selection.

- **MOMBI: Many-objective Meta-Heuristic Based on the R2 Indicator**

MOMBI was introduced by Gomez and Coello Coello [49]. The MOMBI algorithm is based on the R2 indicator which is defined as:

$$R2(A, V, Z^*) = \frac{1}{|V|} \sum_{v \in V} \min_{a \in A} \left\{ \max_{1 \le j \le m} v_j |Z_j^* - a_j| \right\} \quad (20)$$

where A is an individual set, V is a set of weight vectors, and Z^* is used as a reference point which is never dominated by any feasible solution. This algorithm produces a non-dominated sorting scheme based on the utility functions. The main idea is to group solutions that optimize the set of utility functions and gives them the first rank. Then, those solutions will be removed and a second rank will be identified in the same manner. The process will continue until all the population members will be ranked. We notice that MOMBI uses the non-dominated sorting scheme without using the usual Pareto dominance. The MOMBI algorithm is described as follows. MOMBI begins by generating a population P randomly. Then, we obtain the objective function values, the ideal and the nadir point, and the R2-ranking of all P members. After that, a binary tournament selection using the rank of the solutions and the mutation and crossover operators are performed to create an offspring population Q. Next, the reference points are updated with the minimum and maximum objective function values and the population R which is the union of both P and Q populations is ranked using the R2 indicator. In order to reduce the population, MOMBI selects the best N individuals according to their ranks. The experimental results show that MOMBI outperforms MOEA/D [50] in most cases. This algorithm performs well when dealing with many-objective. However, its main weakness is its high computational cost.

- **Other Indicator-Based Approaches**

Indicator-based approaches are yet another direct way to solve MaOPs [51]. In fact, in an indicator-based algorithm, an indicator is not only used to evaluate the obtained approximation set according to the indicator but also indicator values are used to guide the search process. Although, an emerging trend is the use of a quality indicator to solve a MaOP. We identify two indicators that have been applied

Table 5 Comparison of indicator-based approaches for MaOPs: MaxObjs means maximum number of objectives

Classes	Algorithms	References	MaxObjs	Test problems
Hypervolume-based approaches	IBEA	Zitzler and Künzli [43]	4	EXPO
	SMS-EMOA	Wagner and Neumann [47]	20	DTLZ
				WFG
				LZ
	HypE	Bader and Zitzler [52]	50	DTLZ
				WFG
				MKP
R2 indicator-based approaches	AGE	Bringmann et al. [48]	20	DTLZ
	AGE-II	Wagner and Neumann [47]	20	DTLZ
	MOMBI	Gomez and Coello Coello [49]	8	DTLZ
				WFG
	R2-MOGA	Manriquez et al. [54]	10	DTLZ
	R2MODE	Manriquez et al. [54]	10	DTLZ

by most indicator-based approaches for MaOPs: Hypervolume indicator and R2 indicator (cf. Table 5). In fact, tow main issues arise when using the hypervolume indicator to solve MaOPs. First, the computational cost of the hypervolume value is high. Second, the hypervolume might not be appropriate when the DM aims to find a uniform spread optimal set.

In order to deal with the high computational cost of computing the exact hypervolume values, Bader and Zitzler [52] introduced the Hypervolume Estimation Algorithm (HypE) where they used a Monte Carlo algorithm [53] in order to approximate the exact hypervolume values. In this algorithm, the non-dominated solutions are compared according to their hypervolume-based fitness values. Specifically, HypE uses an environmental selection to create a new population from the best solutions in the union set of the parent and offspring populations and estimate the hypervolume value by sampling solutions in different fronts. The experimental results showed that HypE achieved competitive performance in terms of the average hypervolume on a number of test problems with up to 50 objectives. Manriquez et al. [54] proposed two R2-indicator-based approaches which are: R2-MOGA and R2MODE. Those latter present a modified version of Goldberg's non-dominated sorting method. The obtained results on DTLZ with up to 10 objectives indicate that these algorithms can outperform SMS-EMOA in term of computational time.

2.5 Decomposition-Based Approaches

- **MOGLS: Multi-objective Genetic Local Search**

MOGLS was first proposed by Ishibuchi and Murata [55] and improved by Jaszkiewicz [56]. In fact, the Genetic Local Search (GLS) is a metaheuristic that hybridizes recombination operators with local search or with other local improvement heuristics. The basic idea of MOGLS is to transform the original MaOP into a simultaneous optimization of a collection of weighted Tchebycheff functions or weighted sum functions. At each iteration, the algorithm generates a random weight vector to evaluate the current population and uses an external population to store the non-dominated solutions. The Jaszkiewicz's MOGLS can be described as follows:

- **Step 1**: An initialization step is performed to initialize a set of current solutions *CS* with *S* solutions, a vector $z = (z_1, z_2, \ldots, z_m)^T$ where z_i is the largest value found so far for the objective f_i, and an external population *EP* to store the non-dominated solutions of *CS*;
- **Step 2**: Then, the external population *EP* is updated as follows:

1. A randomly weight vector w is generated, k (i.e., the size of temporary elite population) best solutions with regard to the used scalarizing function are selected to form a temporary elite population T, and a new solution y is generated by applying the genetic operators to two randomly chosen solutions from T;
2. A solution y' is generated by applying a local improvement heuristic to y;
3. The vector z is updated: For each $j = 1, \ldots, m$, if $z_j < f_j(y')$, then set $z_j = f_j(y')$. This step is performed only in the case where the Tchebycheff approach is used, this step is removed otherwise;
4. The solution y' is added to *CS*, if y' is better than the worst solution in T with regard to the used scalarizing function and different from the solutions in T with regard to the m real-valued objective functions. In the case where the size of *CS* is larger than $K \times S$ the oldest solution is deleted from *CS*.
5. All the solutions in *EP* that are dominated by y' are removed and y' is added to *EP* if there is no solutions that dominate it.

The above described steps are repeated until a stopping criterion is satisfied. The experimental results have shown that MOGLS may work well on MaOPs. However, the use of the recombination operator and the appropriate selection of the solutions for recombination influence the performance of MOGLS. Moreover, as reported in [50], the upper bound of the size of *CS* which is equal to $K \times S$ influences the space complexity of MOGLS.

- **MOEA/D: Multi-objective Evolutionary Algorithm Based on Decomposition**

MOEA/D is one of the most popular decomposition-based algorithm proposed by Zhang and Li [50]. MOEA/D decomposes the MaOP into N sub-problems (N is the population size) that are optimized simultaneously. It uses a set of well-distributed weight vectors λ_j to cover the whole Pareto front. The algorithm begins by determining a neighborhood of T weight vectors for each λ_j. After that, the population members are assigned to the weight vectors. Thereafter, two solutions from neighboring weight vectors are mated and an offspring solution is created. The offspring solution is then evaluated using a scalarizing function. This generated new solution can also replace several current solutions of its neighboring sub-problems when it outperforms them. Three versions of scalarizing functions are adopted for MOEA/D: (1) weighted sum approach [57], (2) weighted Tchebycheff approach [57], and (3) boundary intersection approach [58, 59]. Ishibuchi et al. [60] studied the relation between the neighborhood size and the performance of MOEA/D in solving many-objective problems. In this work, it was proved that a large replacement neighborhood improves the search ability of MOEA/D in the objective space. However, a small replacement and mating neighborhood are beneficial to maintain the diversity. MOEA/D has demonstrated very interesting results on several MaOPs. However, its main shortcoming is the degradation of diversity and solution distribution when tackling scaled problems.

- **NSGA-III: Non-dominated Sorting Genetic Algorithm III**

Deb and Jain [61] proposed NSGA-III which remains similar to the NSGA-II algorithm with some changes in its selection mechanism. The general principle of this MaOEA can be described as follows. Differently to MOEA/D, NSGA-III makes the decomposition based on a set of well-distributed reference points. Afterwords, a randomly parent population P_t with N individuals is generated. The following steps are iterated until the termination criterion is satisfied. The algorithm begins by creating an offspring population Q_t with N individuals obtained by applying genetic operators to P_t. Thereafter, the two populations P_t and Q_t are merged with each other to form a new population R_t of size $2N$. After that, the combined population R_t is sorted into several fronts using the non-dominated sorting as done in NSGA-II. Then, a new population S_t is constructed starting from the first front F_1 until the size of the population S_t becomes equal to N or for the first time greater than N. Let us suppose that the last accepted level is the lth level. Therefore, all solutions from level $(l + 1)$ onwards are rejected. In most cases, the last front F_l is accepted partially. NSGA-II uses a niching strategy to choose individuals from the last front which are situated in the least crowding regions in F_l. However, the crowding distance is not well-suited for MaOPs. For this reason, the selection mechanism was modified in NSGA-III. Figure 3 illustrates the two mechanisms used in (a) NSGA-II and (b) NSGA-III to maintain diversity among solutions. The principle of the selection mechanism is as follows. It begins by normalizing the population members and the supplied reference points. Then, it calculates the perpendicular distance between a solution in S_t and each of the reference lines that join the ideal point with the reference points. So that, each individual in S_t is associated with the reference point having the minimum perpendicular distance. Thereafter, a niche preservation operation is performed and it can be summarized by this two following steps:

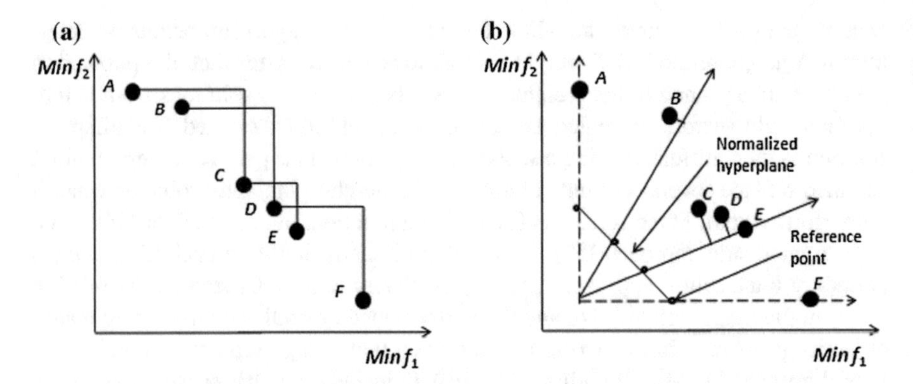

Fig. 3 Illustration of working principles of **a** NSGA-II versus **b** NSGA-III (inspired by [62])

- **Step 1**: Count the number of individuals from $P_{t+1} = S_t/F_l$ that are associated with each reference point;
- **Step 2**: Define a reference point set that contains the reference points having the minimum niche count ρ. If this set contains more than one point, we choose one of them at random.

Hence, four scenarios are identified which are detailed in [61]. After that, we update the different niche counts. It should be noted that this procedure is repeated until the population size of P_{t+1} becomes equal to N. NSGA-III has demonstrated very good results on problems involving up to 15 objectives. The major advantage of NSGA-III is its ability to find well-converged and well-diversified solutions. Another advantage is that it does not require any additional parameters to be set such in MOEA/D.

- **DBEA-Eps: Decomposition Based Evolutionary Algorithm for Many-objective Optimization with Systematic Sampling and Adaptive Epsilon Control**

Asafuddoula et al. [63] proposed a decomposition-based algorithm that generates a structured set of reference points, that uses an adaptive epsilon comparison to manage the balance between the convergence and the diversity, and that adopts an adaptive epsilon formulation to deal with constraints. DBEA-Eps begins with a generation of a set of reference points using the normal boundary intersection method (NBI). Thereafter, the neighborhood of each reference point (i.e., T closest reference points computed based on a Euclidean distance amongst them) is determined. Similarly to NSGA-III, DBEA-Eps normalizes the population based on intercepts calculated using M extreme points of the non-dominated set and computes the same two distance measures d_1 and d_2 used in NSGA-III to control diversity and convergence of the algorithm. Figure 4 illustrates the two distance measures d_1 and d_2 in a two objectives minimization problem. It also uses a mating partner selection to select a parent from the neighborhood of the current solution P_i with a given mating probability δ and a method of recombination using information from neighboring sub-problems. In order

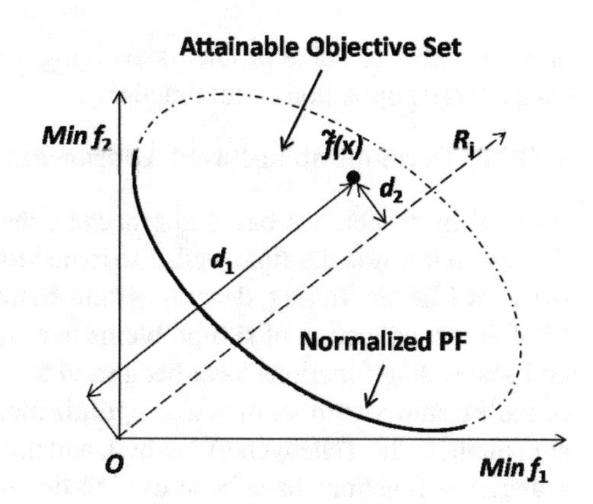

Fig. 4 Illustration of the distance measures d_1 and d_2 with respect to a reference direction. From Ref. [63]

to manage the balance between convergence and diversity, the authors proposed to use an adaptive epsilon comparison, where a child solution replaces a single parent based on the following equation:

$$(d_1, d_2) < \varepsilon_{CD}(d_1, d_2) \Rightarrow \begin{cases} d_1 < d_2, & \text{if } d_2,\ d_2 < \varepsilon_{CD} \\ d_1 < d_2, & \text{if } d_2 = d_2 \\ d_1 < d_2, & \text{otherwise} \end{cases} \tag{21}$$

where d_{2i} is the d_2 measure of the i-th individual, W is the number of reference points, and the average deviation ε_{CD} is defined as follows:

$$\varepsilon_{CD} == \frac{\sum_{i=1}^{W} d_{2i}}{W} \tag{22}$$

In this work, an epsilon level comparison is used to compare the solutions. The DBEA-Eps has demonstrated its outperformance on the DTLZ1 and DTLZ2 problems and on the three constrained engineering design optimization problems with three to seven constraints (car side impact [64], water resource management [65], and a general aviation aircraft design problem [66]). Thus, it is able to deal with unconstrained and constrained MaOPs. However, the performance is dependent on the choice of a number of parameters and several adaptive rules.

The same authors [67] have proposed the improved decomposition based evolutionary algorithm (I-DBEA) which is a modified version of DBEA-Eps. I-DBEA eliminates the use of the neighborhood size T and the mating probability δ such that the entire population is considered as a neighborhood and a first encounter replacement strategy has been adopted. Comparisons between solutions were based on an adaptive epsilon level of d_2. However, in I-DBEA, a simple precedence rule is used, where d_2 has a precedence over d_1. In the proposed algorithm, a corner-sort is used to identify M extreme points that are used to create the hyperplane and to compute the intercepts. The experimental results indicate that I-DBEA is able to deal with unconstrained and constrained MaOPs. However, as noted by the authors, this approach is not suitable to solve problems evolving a large number of reference directions (i.e., a large population is not practical).

- **Other Decomposition-Based Approaches**

Inspired by Preference based approaches, the researchers have proposed to direct the search towards multiple well-distributed ROIs in order to cover the whole Pareto front for MaOPs. In fact, decomposition-based approaches decompose the original MOP into a collection of sub-problems that will be simultaneously optimized. Several scalarizing functions have been used to convert the problem of approximation of the PF into a number of scalar optimization sub-problems such as the weighted sum method, the Tchebycheff method, and the boundary intersection method. Those scalarizing functions have been used to decompose the problem into single objective sub-problems which are defined with the help of weight vectors (Miettinen and Mäkelä [68]). Li et al. [69] proposed MOEA/DD which presents a unified paradigm that combines dominance and decomposition-based approaches for many-objective optimization to balance between convergence and diversity. MOEA/DD uses an update procedure that depends on Pareto dominance, local density estimation, and scalarizing functions, sequentially. The authors have also proposed a modified version of MOEA/DD called C-MOEA/DD to solve constrained problems. The performance of the two algorithms was investigated on a set of unconstrained benchmark problems with up to fifteen objectives and on a number of constrained optimization problems. The obtained results have demonstrated the outperformance of both algorithms in solving problems with a high number of objectives. However, MOEA/DD is sensitive to the two parameters T and δ which represent the neighborhood size and the probability of selecting mating parents from neighboring sub-regions, respectively. Different MOEA/D variants have been proposed in the literature to tackle MaOPs such as MOEA/D-DRA (Zhang et al. [70]) and UMOEA/D (Tan et al. [71]). Yuan et al. [72] proposed the θ-NSGA-III witch is an improved version of NSGA-III, but the main difference between the two algorithms is that the θ-NSGA-III replaces the Pareto dominance used in NSGA-III with a new dominance relation which is called the θ-dominance. θ-NSGA-III outperforms MOEA/D and NSGA-III in terms of convergence. However, it was proved that this algorithm is insensitive to the parameter θ. Elarbi et al. [73] proposed a new dominance relation called TSD-dominance to deal with MaOPs. The TSD-NSGA-II represents a new many-objective version of NSGA-II where the Pareto dominance is replaced by the TSD-dominance. TSD-NSGA-II was found to be highly competitive in dealing with constrained and unconstrained problems. However, MaOPs involving the characteristics of DTLZ6-7 represent the limits of TSD-NSGA-II. Table 6 provides a comparison of some of the most prominent decomposition-based approaches for MaOPs. From the different discussed works in this chapter, we remark that the choice of a specific scalarizing function to use influences the performance of the decomposition-based algorithm

Table 6 Comparison of decomposition-based approaches for MaOPs: MaxObjs means maximum number of objectives, CP means constrained problems, UP means unconstrained problems, WV means weight vectors, RP means reference points

Algorithms	References	MaxObjs	CP	UP	WV	RP
MOGLS	Jaszkiewicz [56]	4	X	–	X	–
MOEA/D	Zhang and Li [50]	4	X	X	X	–
NSGA-III	Deb and Jain [61]	15	–	X	–	X
DBEA-Eps	Asafuddoula et al. [63]	15	X	X	–	X
I-DBEA	Asafuddoula et al. [67]	15	X	X	–	X
MOEA/DD	Li et al. [69]	15	X	X	X	–
MOEA/D-DRA	Zhang et al. [70]	5	–	X	X	–
UMOEA/D	Tan et al. [71]	5	X	X	X	–
θ-NSGA-III	Yuan et al. [72]	20	–	X	–	X
TSD-NSGA-II	Elarbi et al. [73]	20	–	X	–	X

[74]. Moreover, different methods have been used to generate a set of weight vectors (i.e., reference points) such as the systematic approach [58] and the on-the-fly weighting vector generating method [75]. However, how to configure the weight vectors is still a big challenge for decomposition-based algorithms, since those latter dramatically affect the diversity performance.

3 Performance Assessment of MaOEAs

3.1 Test Problems and Statistical Analysis

Several test problems have been used to investigate MaOEAs capabilities in approximating the Pareto front. In the literature, among the most used test function suites we find: (1) the scalable DTLZ (Deb-Thiele-Laumans- Zitzler) suite and the scalable WFG (Walking Fish Group) Toolkit. MaOEAs have also been used in some combinatorial problems such as knapsack problems. Recently, a number of scalable constrained test problems having three up to 15 objectives have been introduced [64]. Those latter are characterized with various types of difficulties to an algorithm. Table 7 summarizes the above mentioned test problems in this chapter.

Many existing works use the median and the interquartile range values obtained by a specific performance metric in order to compare the performance of different algorithms. However, each algorithm can behave differently from one run to another. Hence, the use of a statistical testing approach is necessary. Firstly, we should check whether the obtained results are normally distributed or not using the Kolmogorov–Smirnov test. If data follow a normal distribution, we can use the t-test when comparing between two algorithms and the ANOVA one if the comparison

Table 7 Summary of the mentioned test problems (inspired by [28])

Test problems	References	Remarks
DTLZ	Deb et al. [21]	Scalable problem
WFG	Huband et al. [22]	Scalable problem
MKP	Zitzler et al. [9]	Multi-objective 0/1 Knapsack Problem
NRP	Burke et al. [18]	Nurse Restoring Problem
DTLZ5(I,M)	Deb et al. [19]	Scalable problem
DTLZ2$_{BZ}$	Brockhoff et al. [77]	Modified version of DTLZ2
LPMS	Miettinen et al. [78]	Locating a pollution monitoring station
EXPO	Thiele et al. [79]	A network processor application comprising problem
LZ	Li et al. [80]	Continuous test problems
Car side impact	Jain and Deb [64]	Engineering constrained problem
Water resource management	Ray et al. [65]	Engineering constrained problem
General aviation aircraft design	Hadka et al. [66]	Engineering constrained problem

involves more than two algorithms. To avoid verifying data normality, we can directly use the Wilcoxon test and the Kruskal-Wallis as non-parametric alternatives of the t-test and the ANOVA one respectively. For more details about statistical testing in evolutionary computation, the reader could refer to [76].

3.2 Performance Metrics

In the literature, fewer are the performance metrics that have been used to evaluate the performance of MaOEAs. The most common used performance metrics are described in this subsection.

- **The Inverted Generational Distance (IGD)**

The IGD is a performance metric that measures the distance between the true Pareto front and the closest individual in an approximation set. It is expressed as follows [81]:

$$I_{IGD} = \frac{(\sum_{i=1}^{PF^*} d_i^q)^{\frac{1}{q}}}{PF^*} \quad (23)$$

where d_i^q is the Euclidean distance between an individual from the Pareto front PF^* to its nearest individual in the approximation set and $q = 2$. In fact, the lower are the I_{IGD} values, the better are the obtained sets. Moreover, the IGD can measure both convergence and diversity. The IGD metric requires a reference true Pareto front in the calculation. However, it is difficult to determine the reference true Pareto front of MaOPs. Thus, an exact method to generate a set of uniformly well-spread points over the true Pareto front has been proposed for the DTLZ1-DTLZ4 test problems [69]. This method locates exactly the intersecting points of the reference points generated by the algorithm and the Pareto-optimal surface since the exact Pareto-optimal surfaces of DTLZ1 to DTLZ4 are known a priori. For DTLZ1, given a reference point $r = (r_1, \ldots, r_M)^T$, the i-th objective function of a Pareto-optimal solution x^* is computed as follows:

$$f_i(x^*) = 0.5 \times \frac{r_i}{\sum_{j=1}^{M} r_j} \tag{24}$$

As for DTLZ2 to DTLZ4, given a reference point $r = (r_1, \ldots, r_M)^T$, the i-th objective function of a Pareto-optimal solution x^* is computed as follows:

$$f_i(x^*) = \frac{r_i}{\sqrt{\sum_{j=1}^{M} r_j^2}} \tag{25}$$

- **The Generational Distance (GD)**

The GD metric evaluates an average distance of an approximation set P from the true Pareto front PF^* [82]. It is defined as follows:

$$I_{GD} = \frac{\sqrt{\sum_{i=1}^{P} d_i^2}}{|P|} \tag{26}$$

where d_i is the Euclidean distance between the solution $i \in P$ and the nearest member of PF^*. A value of $I_{GD} = 0$ indicates that all the individuals of the approximation set P are in the true Pareto front PF^*. This metric evaluate only the convergence of an algorithm.

- **The Hypervolume (HV)**

The HV indicator is a unary indicator that calculates the volume of the hypercube dominated by an approximation set. This indicator can be expressed as follows:

$$I_{HV} = \bigcup_i S(i) | i \in PF^* \tag{27}$$

where $S(i)$ is the hypercube bounded by a solution i and a reference point. In fact, the choice of the reference point is important because it influences the outcome of this metric. The reference point can be constructed by the worst objective function values. This measure captures both convergence and diversity. A large value of the HV metric is desirable.The main drawback of this metric is the high computational cost needed to compute the exact HV [53].

- **The Spread** (Δ)

The Δ metric measures the deviation among neighboring solutions in the non-dominated solution set P furnished by the MOEA [83]. Analytically, it is stated as follows:

$$I_{\Delta} = \sum_{i=1}^{|P|} \frac{|d_i - \overline{d}|}{|P|} \tag{28}$$

where d_i is the Euclidean distance between two neighbor solutions in P and $\overline{d}$ is the average of these distances. In fact, a smaller value of Δ indicates that the algorithm is able to find a diverse set of non-dominated solutions.

4 Conclusion and Future Research Paths

In this chapter, we have first described the related issues that MOEAs encounter when the dimensionality of the objective space increases. Then, we have surveyed the most prominent MaOEAs. We have proposed to classify the existing MaOEAs into five classes: Preference ordering relation-based approaches, objective reduction-based approaches, preference incorporation-based approaches, indicator-based approaches, and decomposition-based approaches.

The preference ordering relation-based approach aims to propose a preference relation that induce a finer order than that induced by the Pareto dominance relation. Hence, the non-dominated solutions are further ranked using this relation. This method has the ability to increase the selection pressure towards the Pareto front. However, it decreases the diversity of the solutions. Thus, it will be interesting to propose new flexible selection methods and new diversity mechanisms for preference ordering relation-based approaches.

The objective reduction-based approach finds the relevant objectives and eliminates the redundant objectives that are not essential to describe the Pareto optimal front. In other words, it identifies the non-conflicting objectives and discards them to reduce the number of objectives of the MaOPs during the search process. Two reduction methods can be identified: (1) the offline dimensionality reduction method and (2) the online dimensionality reduction method. The main advantage of this approach is that it reduces the computational cost of the MaOEAs. However, for MaOps with non-redundant objectives this approach may fail to reduce the number of objectives.

The preference incorporation-based approach exploits the DM preferences in order to differentiate between Pareto equivalent solutions. It focuses the search process on a specific region of the Pareto front. Preference incorporation-based methods can be classified into the three following subclasses: (1) priori preference-based approaches, (2) interactive preference-based approaches, and (3) posteriori preference-based approaches. In fact, one of the issues that arises when using the a posteriori preference-based approach is that it may obtain a high number of solutions that the DM is not interested in [28].

The indicator-based approach transforms the MaOP into the problem of optimizing an indicator by evaluating the solutions using a performance metric. The high computational cost of the hypervolume calculation represents a difficulty for this approach when dealing with high dimensional objective space. Hence, it will be interesting to propose new indicators that are well-adapted for MaOPs.

The decomposition-based approach decomposes the problem into several sub-problems that will be simultaneously optimized using scalarizing functions. Actually, decomposition is the most successful approach to solve MaOPs. Both the scalarizing function and the method used to generate a structured set of reference points (or weight vectors) influence the performance of a decomposition-based algorithm. For this reason, more future research are needed on proposing new methods for generating weight vectors.

References

1. Bechikh, S., Chaabani, A., Said, L.B.: An efficient chemical reaction optimization algorithm for multiobjective optimization. IEEE Trans. Cybern. **45**(10), 2051–2064 (2015)
2. Deb, K., Pratap, A., Agarwal, S., Meyarivan, T.: A fast and elitist multiobjective genetic algorithm: Nsga-ii. IEEE Trans. Evolut. Comput. **6**(2), 182–197 (2002)
3. Zitzler, E., Laumanns, M., Thiele, L., Zitzler, E., Zitzler, E., Thiele, L., Thiele, L.: Spea 2: Improving the strength pareto evolutionary algorithm (2001)
4. Sülflow, A., Drechsler, N., Drechsler, R.: Robust multi-objective optimization in high dimensional spaces. In: Evolutionary Multi-criterion Optimization, pp. 715–726. Springer, Heidelberg (2007)
5. Kasprzyk, J.R., Reed, P.M., Kirsch, B.R., Characklis, G.W.: Managing population and drought risks using many-objective water portfolio planning under uncertainty. Water Resour. Res. **45**(12) (2009)
6. Garza-Fabre, M., Pulido, G. Coello, C.A.C.: Ranking methods for many-objective optimization. In: MICAI 2009: Advances in Artificial Intelligence, pp. 633–645. Springer, Heidelberg (2009)
7. Jaimes, A.L.: Técnicas para resolver problemas de optimizacíon con muchas funciones objetivo usando algoritmos evolutivos. Ph.D. thesis (2011)
8. Sato, H., Aguirre, H.E., Tanaka, K.: Controlling dominance area of solutions and its impact on the performance of moeas. In: Evolutionary Multi-criterion Optimization, pp. 5–20. Springer, Heidelberg (2007)
9. Zitzler, E., Thiele, L.: Multiobjective optimization using evolutionary algorithms-a comparative case study. In: Parallel Problem Solving from Nature-PPSN V, pp. 292–301. Springer, Heidelberg (1998)
10. Farina, M., Amato, P.: On the optimal solution definition for many-criteria optimization problems. In: Proceedings of the NAFIPS-FLINT International Conference, pp. 233–238 (2002)
11. Bentley, P.J., Wakefield, J.P.: Finding acceptable solutions in the pareto-optimal range using multiobjective genetic algorithms. In: Soft Computing in Engineering Design and Manufacturing, pp. 231–240. Springer, Heidelberg (1998)

12. Li, M., Zheng, J., Li, K., Yuan, Q., Shen, R.: Enhancing diversity for average ranking method in evolutionary many-objective optimization. In: Parallel Problem Solving from Nature, PPSN XI, pp. 647–656. Springer, Heidelberg (2010)
13. Corne, D.W., Knowles, J.D.: Techniques for highly multiobjective optimisation: some nondominated points are better than others. In: Proceedings of the 9th Annual Conference on Genetic and Evolutionary Computation, pp. 773–780. ACM, New York (2007)
14. Drechsler, N., Drechsler, R., Becker, B.: Multi-objective optimisation based on relation favour. In: Evolutionary Multi-criterion Optimization, pp. 154–166. Springer, Heidelberg (2001)
15. Drechsler, N., Drechsler, R., Becker, B.: Multi-objective optimization in evolutionary algorithms using satisfiability classes. In: Computational Intelligence, pp. 108–117. Springer, Heidelberg (1999)
16. Di Pierro, F., Khu, S.-T., Savic, D.A.: An investigation on preference order ranking scheme for multiobjective evolutionary optimization. IEEE Trans. Evolut. Comput. **11**(1), 17–45 (2007)
17. Das, I.: A preference ordering among various pareto optimal alternatives. Struct. Optim. **18**(1), 30–35 (1999)
18. Burke, E.K., De Causmaecker, P., Berghe, G.V., Van Landeghem, H.: The state of the art of nurse rostering. J. Sched. **7**(6), 441–499 (2004)
19. Deb, K., Saxena, D.: Searching for pareto-optimal solutions through dimensionality reduction for certain large-dimensional multi-objective optimization problems In: Proceedings of the World Congress on Computational Intelligence (WCCI-2006), pp. 3352–3360 (2006)
20. Singh, H.K., Isaacs, A., Ray, T.: A pareto corner search evolutionary algorithm and dimensionality reduction in many-objective optimization problems. IEEE Trans. Evolut. Comput. **15**(4), 539–556 (2011)
21. Deb, K., Thiele, L., Laumanns, M., Zitzler, E.: Scalable multi-objective optimization test problems. In: Proceedings of the 2002 Congress on Evolutionary Computation, 2002. CEC'02, vol. 1, pp. 825–830. IEEE, New York (2002)
22. Huband, S., Hingston, P., Barone, L., While, L.: A review of multiobjective test problems and a scalable test problem toolkit. IEEE Trans. Evolut. Comput. **10**(5), 477–506 (2006)
23. Jaimes, A.L, Coello, C.A.C., Chakraborty, D.: Objective reduction using a feature selection technique. In: Proceedings of the 10th annual conference on Genetic and evolutionary computation, pp. 673–680. ACM, New York (2008)
24. Mitra, P., Murthy, C., Pal, S.K.: Unsupervised feature selection using feature similarity. IEEE Trans. Pattern Anal. Mach. Intell. **24**(3), 301–312 (2002)
25. Brockhoff, D., Saxena, D.K., Deb, K., Zitzler, E.: On handling a large number of objectives a posteriori and during optimization. In: Multiobjective Problem Solving from Nature, pp. 377–403. Springer, Heidelberg (2008)
26. Saxena, D.K., Duro, J.A., Tiwari, A., Deb, K., Zhang, Q.: Objective reduction in many-objective optimization: linear and nonlinear algorithms. IEEE Trans. Evolut. Comput. **17**(1), 77–99 (2013)
27. Brockhoff, D., Zitzler, E.: Are all objectives necessary? on dimensionality reduction in evolutionary multiobjective optimization. In: Parallel Problem Solving from Nature-PPSN IX, pp. 533–542. Springer, Heidelberg (2006)
28. Li, B., Li, J., Tang, K., Yao, X.: Many-objective evolutionary algorithms: a survey. ACM Comput. Surv. (CSUR) **48**(1), 13 (2015)
29. Deb, K., Sundar, J.: Reference point based multi-objective optimization using evolutionary algorithms. Int. J. Comput. Intell. Res. **2**, 273–286 (2006)
30. Thiele, L., Miettinen, K., Korhonen, P.J., Molina, J.: A preference-based evolutionary algorithm for multi-objective optimization. Evolut. Comput. **17**(3), 411–436 (2009)
31. Said, L.B, Bechikh, S., Ghédira, K.: The r-dominance: a new dominance relation for interactive evolutionary multicriteria decision making. IEEE Trans. Evolut. Comput., **14**(5), 801–818 (2010)
32. Zitzler, E., Deb, K., Thiele, L.: Comparison of multiobjective evolutionary algorithms: empirical results. Evolut. Comput. **8**(2), 173–195 (2000)

33. Wang, R., Purshouse, R.C., Fleming, P.J.: Preference-inspired coevolutionary algorithms for many-objective optimization. IEEE Trans. Evolut. Comput. **17**(4), 474–494 (2013)
34. Bechikh, S., Kessentini, M., Said, L.B., Ghédira, K.: Chapter four-preference incorporation in evolutionary multiobjective optimization: a survey of the state-of-the-art. Adv. Comput. **98**, 141–207 (2015)
35. Bechikh, S.: Incorporating decision maker's preference information in evolutionary multi-objective optimization. Ph.D. thesis, University of Tunis, ISG-Tunis, Tunisia (2013)
36. Bechikh, S., Said, L.B., Ghédira, K.: Negotiating decision makers' reference points for group preference-based evolutionary multi-objective optimization. In: 2011 11th International Conference on Hybrid Intelligent Systems (HIS), pp. 377–382. IEEE, New York (2011)
37. Bechikh, S., Said, L.B., Ghédira, K.: Group preference based evolutionary multi-objective optimization with nonequally important decision makers: Application to the portfolio selection problem. Int. J. Comput. Inf. Syst. Ind. Manag. Appl. **5**(278–288), 71 (2013)
38. Kalboussi, S., Bechikh, S., Kessentini, M., Said, L.B.: Preference-based many-objective evolutionary testing generates harder test cases for autonomous agents. In: Search Based Software Engineering, pp. 245–250. Springer, Heidelberg (2013)
39. Dunwei, G., Gengxing, W., Xiaoyan, S.: Set-based genetic algorithms for solving many-objective optimization problems. In: 2013 13th UK Workshop on Computational Intelligence (UKCI), pp. 96–103 (2013)
40. Deb, K., Kumar, A.: Interactive evolutionary multi-objective optimization and decision-making using reference direction method. In: Proceedings of the 9th Annual Conference on Genetic and Evolutionary Computation, pp. 781–788. ACM, New York (2007)
41. Wang, R., Purshouse, R.C., Fleming, P.J.: On finding well-spread pareto optimal solutions by preference-inspired co-evolutionary algorithm. In: Proceedings of the 15th Annual Conference on Genetic and Evolutionary Computation, pp. 695–702. ACM, New York (2013)
42. Wang, R., Purshouse, R.C., Fleming, P.J.: Preference-inspired co-evolutionary algorithm using weights for many-objective optimization. In: Proceedings of the 15th Annual Conference Companion on Genetic and Evolutionary Computation, pp. 101–102. ACM, New York (2013)
43. Zitzler, E., Künzli, S.: Indicator-based selection in multiobjective search. In: Parallel Problem Solving from Nature-PPSN VIII, pp. 832–842. Springer, Heidelberg (2004)
44. Basseur, M., Burke, E.K.: Indicator-based multi-objective local search. In: IEEE Congress on Evolutionary Computation, 2007. CEC 2007, pp. 3100–3107. IEEE, New York (2007)
45. Wagner, T., Beume, N., Naujoks, B.: Pareto-, aggregation-, and indicator-based methods in many-objective optimization. In: Evolutionary Multi-criterion Optimization, pp. 742–756. Springer, Heidelberg (2007)
46. Emmerich, M., Beume, N., Naujoks, B.: An emo algorithm using the hypervolume measure as selection criterion. In: Evolutionary Multi-Criterion Optimization, pp. 62–76. Springer, Heidelberg (2005)
47. Wagner, M., Neumann, F.: A fast approximation-guided evolutionary multi-objective algorithm. In: Proceedings of the 15th Annual Conference on Genetic and Evolutionary Computation, pp. 687–694. ACM, New York (2013)
48. Bringmann, K., Friedrich, T., Neumann, F., Wagner, M.: Approximation-guided evolutionary multi-objective optimization. IJCAI Proc. Int. Joint Conf. Artif. Intell. **22**, 1198–1203 (2011)
49. Gomez, R.H., Coello, C.C.: Mombi: A new metaheuristic for many-objective optimization based on the r2 indicator. In: 2013 IEEE Congress on Evolutionary Computation (CEC), pp. 2488–2495. IEEE, New York (2013)
50. Zhang, Q., Li, H.: Moea/d: A multiobjective evolutionary algorithm based on decomposition. IEEE Trans. Evolut. Comput. **11**(6), 712–731 (2007)
51. Azzouz, N., Bechikh, S., Said, L.B.: Steady state ibea assisted by mlp neural networks for expensive multi-objective optimization problems. In: Proceedings of the 2014 Conference on Genetic and Evolutionary Computation, pp. 581–588. ACM, New York (2014)
52. Bader, J., Zitzler, E.: Hype: an algorithm for fast hypervolume-based many-objective optimization. Evolut. Comput. **19**(1), 45–76 (2011)

53. Bader, J., Deb, K., Zitzler, E.: Faster hypervolume-based search using monte carlo sampling. In: Multiple Criteria Decision Making for Sustainable Energy and Transportation Systems, pp. 313–326. Springer, Heidelberg (2010)
54. Diaz-Manriquez, A., Toscano-Pulido, G., Coello, C.A.C., Landa-Becerra, R.: A ranking method based on the r2 indicator for many-objective optimization. In: 2013 IEEE Congress on Evolutionary Computation (CEC), pp. 1523–1530. IEEE, New York (2013)
55. Ishibuchi, H., Murata, T.: A multi-objective genetic local search algorithm and its application to flowshop scheduling. IEEE Trans. Syst. Man Cybern. Part C Appl. Rev. **28**(3), 392–403 (1998)
56. Jaszkiewicz, A.: On the performance of multiple-objective genetic local search on the 0/1 knapsack problem-a comparative experiment. IEEE Trans. Evolut. Comput. **6**(4), 402–412 (2002)
57. Miettinen, K.: Nonlinear Multiobjective Optimization, International Series in Operations Research and Management Science, vol. 12 (1999)
58. Dennis, J., Das, I.: Normal-boundary intersection: a new method for generating pareto optimal points in nonlinear multicriteria optimization problems. SIAM J. Optim. **8**(3), 631–657 (1998)
59. Messac, A., Ismail-Yahaya, A., Mattson, C.A.: The normalized normal constraint method for generating the pareto frontier. Struct. Multidiscip. Optim. **25**(2), 86–98 (2003)
60. Ishibuchi, H., Akedo, N., Nojima, Y.: Relation between neighborhood size and moea/d performance on many-objective problems. In: Evolutionary Multi-Criterion Optimization, pp. 459–474. Springer, Heidelberg (2013)
61. Deb, K., Jain, H.: An evolutionary many-objective optimization algorithm using reference-point-based nondominated sorting approach, part i: Solving problems with box constraints. IEEE Trans. Evolut. Comput. **18**(4), 577–601 (2014)
62. Seada, H., Deb, K.: U-nsga-iii: A unified evolutionary algorithm for single, multiple, and many-objective optimization, COIN report, no. 2014022
63. Asafuddoula, M., Ray, T., Sarker, R.: A decomposition based evolutionary algorithm for many objective optimization with systematic sampling and adaptive epsilon control. In: Evolutionary Multi-Criterion Optimization, pp. 413–427. Springer, Heidelberg (2013)
64. Jain, H., Deb, K.: An evolutionary many-objective optimization algorithm using reference-point based nondominated sorting approach, part ii: handling constraints and extending to an adaptive approach. IEEE Trans. Evolut. Comput. **18**(4), 602–622 (2014)
65. Ray, T., Tai, K., Seow, C.: An evolutionary algorithm for multiobjective optimization. Eng. Optim. **33**(3), 399–424 (2001)
66. Hadka, D., Reed, P.M., Simpson, T.W.: Diagnostic assessment of the borg moea for many-objective product family design problems. In: 2012 IEEE congress on Evolutionary computation (CEC), pp. 1–10. IEEE, New York (2012)
67. Asafuddoula, M., Ray, T., Sarker, R.: A decomposition-based evolutionary algorithm for many objective optimization. IEEE Trans. Evolut. Comput. **19**(3), 445–460 (2015)
68. Miettinen, K., Mäkelä, M.M.: On scalarizing functions in multiobjective optimization. OR spectrum **24**(2), 193–213 (2002)
69. Li, K., Deb, K., Zhang, Q., Kwong, S.: An evolutionary many-objective optimization algorithm based on dominance and decomposition. IEEE Trans. Evolut. Comput. **19**(5), 694–716 (2015)
70. Zhang, Q., Liu, W., Li, H.: The performance of a new version of moea/d on cec09 unconstrained mop test instances. IEEE Congr. Evolut. Comput. **1**, 203–208 (2009)
71. Tan, Y.-Y., Jiao, Y.-C., Li, H., Wang, X.-K.: Moea/d+ uniform design: a new version of moea/d for optimization problems with many objectives. Comput. Oper. Res. **40**(6), 1648–1660 (2013)
72. Yuan, Y., Xu, H., Wang, B.: An improved nsga-iii procedure for evolutionary many-objective optimization. In: Proceedings of the 2014 Conference on Genetic and Evolutionary Computation, pp. 661–668. ACM, New York (2014)
73. Elarbi, M., Bechikh, S., Said, L.B.: Solving many-objective problems using targeted search directions. In: Proceedings of the 31th Annual ACM Symposium on Applied Computing, pp.89–96. ACM, Pisa, Italy (2016)

74. Derbel, B., Brockhoff, D., Liefooghe, A., Verel, S..: On the impact of multiobjective scalarizing functions. In: Parallel Problem Solving from Nature–PPSN XIII, pp. 548–558. Springer, Heidelberg (2014)
75. Hughes, E.J.: Msops-ii: A general-purpose many-objective optimiser. In: IEEE Congress on Evolutionary Computation, 2007. CEC 2007, pp. 3944–3951. IEEE, New York (2007)
76. Derrac, J., García, S., Molina, D., Herrera, F.: A practical tutorial on the use of nonparametric statistical tests as a methodology for comparing evolutionary and swarm intelligence algorithms. Swarm Evolut. Comput. **1**(1), 3–18 (2011)
77. Brockhoff, D., Zitzler, E.: Offline and online objective reduction in evolutionary multiobjective optimization based on objective conflicts. TIK Report **269** (2007)
78. Miettinen, K.: Graphical illustration of pareto optimal solutions. In: Multi-objective Programming and Goal Programming, pp. 197–202. Springer, Heidelberg (2003)
79. Thiele, L., Chakraborty, S., Gries, M., Kunzli, S.: Design space exploration of network processor. Netw. Process. Des. Issues Pract. **1**, 55–89 (2002)
80. Li, H., Zhang, Q.: Multiobjective optimization problems with complicated pareto sets, moea/d and nsga-ii. IEEE Trans. Evolut. Comput. **13**(2), 284–302 (2009)
81. Villalobos, C.A.R., Coello, C.A.C.: A new multi-objective evolutionary algorithm based on a performance assessment indicator. In: Proceedings of the 14th Annual Conference on Genetic and Evolutionary Computation, pp. 505–512. ACM, New York (2012)
82. Van Veldhuizen, D.A., Lamont, G.B.: On measuring multiobjective evolutionary algorithm performance. In: Proceedings of the 2000 Congress on Evolutionary Computation, 2000, vol. 1, pp. 204–211. IEEE, New York (2000)
83. Deb, K.: Multi-objective Optimization Using Evolutionary Algorithms, vol. 16. Wiley, New Jersey (2001)

On the Emerging Notion of Evolutionary Multitasking: A Computational Analog of Cognitive Multitasking

Abhishek Gupta, Bingshui Da, Yuan Yuan and Yew-Soon Ong

Abstract Over the past decades, Evolutionary Computation (EC) has surfaced as a popular paradigm in the domain of computational intelligence for global optimization of complex multimodal functions. The distinctive feature of an Evolutionary Algorithm (EA) is the emergence of powerful implicit parallelism as an offshoot of the simple rules of population-based search. However, despite the known advantages of implicit parallelism, it is interesting to note that EAs have almost exclusively been developed to solve only a single optimization problem at a time; seldom has any effort been made to multitask, i.e., to tackle multiple self-contained optimization problems concurrently using the same population of evolving individuals. To this end, inspired by the remarkable ability of the human brain to perform multiple tasks with apparent simultaneity, we present *evolutionary multitasking* as an intriguing direction for EC research. In particular, the paradigm opens doors to the possibility of autonomously exploiting the underlying complementarities between separate (but possibly similar) optimization exercises through the process of *implicit genetic transfer*, thereby enhancing productivity in decision making processes via accelerated convergence characteristics. Along with the design of an appropriately unified solution representation scheme, we present the outline of a recently proposed algorithmic framework for effective multitasking. Thereafter, the efficacy of the approach is substantiated through a series of practical examples in continuous and discrete optimization that highlight the real-world utility of the paradigm.

A. Gupta (✉) · B. Da · Y. Yuan · Y.-S. Ong
School of Computer Science and Engineering, Nanyang Technological University, Singapore, Singapore
e-mail: abhishekg@ntu.edu.sg

B. Da
e-mail: DA0002UI@e.ntu.edu.sg

Y. Yuan
e-mail: yuanyuan@ntu.edu.sg

Y.-S. Ong
e-mail: asysong@ntu.edu.sg

S. Bechikh et al. (eds.), *Recent Advances in Evolutionary Multi-objective Optimization*, Adaptation, Learning, and Optimization 20,
DOI 10.1007/978-3-319-42978-6_5

Keywords Evolutionary multitasking · Multi-objective optimization · Memetic computation

1 Introduction

One of the most astonishing aspects of human cognition is its ability to manage and execute multiple tasks with what appears to be apparent simultaneity. It is recognized that in this fast-paced, technologically driven world that we live in, the explosion in volume and variety of incoming information streams presents unprecedented opportunity, tendency, and (even) the need to effectively multitask. Merely a fleeting glance at the world around us reveals the ubiquity of supposed cognitive multitasking. From relatively straightforward examples, such as phoning while walking, to more complex ones, such as media multitasking, the human brain has shown notable adaptability to multitask settings. In fact, it is generally acknowledged that multitasking is perhaps the only way to fit in all our priorities into increasingly busy schedules, albeit at the (often tolerable) cost of a marginal drop in the quality of output achieved. Thus, it is not unnatural to expect the pursuit of intelligent systems and algorithms capable of effective multitasking to gain popularity among scientists and engineers who are constantly aiming for enhanced productivity in a world that routinely presents a multiplicity of complex challenges.

It is noted that a major criticism leveled against cognitive multitasking originates from an observed *switching cost* during which the brain attempts to overcome the interference between tasks and adjusts to the new task [1]. Thus, while constantly switching between competing tasks, an individual may often experience slower response times, degraded performance, and/or increased error rates [2]. In this regard, while developing computational analogues of multitasking, it is observed that modern-day computers are in the most part free from any significant switching cost while handling multiple tasks at once. This observation forms grounds for our contention that an artificial (computational) multitasking engine may be capable of retaining many of the advantages of cognitive multitasking, while effectively overcoming its potential perils.

In the field of computational intelligence, Evolutionary Algorithms (EAs) constitute a family of stochastic optimizers that are inspired by Darwinian principles of natural selection [3–5]. The increasing popularity of EAs as a mainstay of optimization in science, operations research, and engineering is largely due to the emergent properties of implicit parallelism of population-based search [6], which circumvents the need for derivative-based techniques that impose continuity and differentiability requirements on objective function landscapes. In fact, it is largely due to the efficient exploitation of implicit parallelism that Multi-objective Evolutionary Algorithms (MOEAs) have rapidly gained in popularity in recent decades, enabling synchronous convergence to a diverse set of near optimal trade-off points [7–9]. Encouraged by this observation, a central goal of the present proposition is to further leverage upon the known power of implicit parallelism, thereby establishing a new niche for EAs that undeniably sets them apart from existing mathematical optimization procedures. In particular, we investigate the potential utility of EAs towards *multitask optimization*, i.e., the solution of multiple self-contained (but possibly similar) optimization tasks at the same time using a single population of evolving individuals. While the proposition bears resembling conceptual motivation to the field of multitask learning [10, 11], it operates from the standpoint of nature-inspired computing, facilitating implicit information exchange across different numerical optimization tasks. To elaborate, we contend that useful inductive biases or some form of knowledge overlap may exist in the evolutionary search of one or more optimization tasks that lie outside the self-contained scope of a particular problem of interest. Neglecting this information, as is typically the case in tabula rasa optimization, may be deemed highly counterproductive, especially given the increasing complexity of real-world problems. In such scenarios, *evolutionary multitasking* provides the scope for autonomously exploiting the complementarities in an implicit manner (through the process of *genetic transfer*), and consequently accelerating convergence characteristics by circumventing several (often impeding) function evaluations [12–14].

For a more detailed illustration of the various notions discussed heretofore, the remainder of this chapter is organized as follows. In Sect. 2, we introduce the preliminaries of multitask optimization. Following [12], we hereafter label the paradigm as *multifactorial optimization* (MFO) in order to emphasize that each task presents an additional *factor* influencing the evolution of a single population. Further, we highlight the key conceptual distinction between multitasking and multi-objective optimization in order to address several queries arising in this regard. In Sect. 3, we present the *Multifactorial Evolutionary Algorithm* (MFEA) from [12], an approach that draws inspiration from bio-cultural models of multifactorial inheritance [15–18]. The means by which the MFEA facilitates knowledge transfer across tasks is also briefly discussed therein. Thereafter, Sect. 4 contains recent case studies for a variety of practical applications of multitasking, including examples in continuous and discrete optimization. In essence, it is reasoned that there exist numerous promising opportunities for MFO in real-world problems, which encourages future research efforts in this direction. Finally, Sect. 5 summarizes the chapter, highlighting important research questions brought to the table by the promising future prospects of multitask optimization.

2 Preliminaries

Consider a hypothetical situation wherein K self-contained optimization tasks are to be performed concurrently. Without loss of generality, all tasks are assumed to be minimization problems. The j-th task, denoted T_j, is considered to have a search space X_j on which the objective function is defined as $F_j : \boldsymbol{X}_j \rightarrow \mathbb{R}$. In addition, each task may be constrained by several equality and/or inequality conditions that must be satisfied for a solution to be considered feasible. In such a setting, we define MFO as an evolutionary multitasking paradigm that aims to simultaneously navigate the design space of all tasks, constantly building on the implicit parallelism of population-based search so as to rapidly deduce $\{\boldsymbol{x}_1, \boldsymbol{x}_2, \ldots, \boldsymbol{x}_{K-1}, \boldsymbol{x}_K\} =$

$argmin\{F_1(\boldsymbol{x}), F_2(\boldsymbol{x}), \ldots, F_{K-1}(\boldsymbol{x}), F_K(\boldsymbol{x})\}$, where $\boldsymbol{x}_j$ is a feasible solution in $\boldsymbol{X}_j$. As suggested by the nomenclature, herein each F_j is treated as an additional factor influencing the evolution of a single population of individuals. For this reason, the composite problem may also be referred to as a K-factorial problem.

While designing evolutionary solvers for MFO, it is necessary to formulate a general technique for comparing population members in a multitasking environment. To this end, we first define a set of properties for every individual p_i, where $i \in \{1, 2, |P|\}$, in a population P. Note that the individuals are encoded in a unified search space $\boldsymbol{Y}$ encompassing $\boldsymbol{X}_1, \boldsymbol{X}_2, \ldots, \boldsymbol{X}_K$, and can be decoded into a task-specific solution representation with respect to each of the K optimization tasks. The decoded form of p_i can thus be written as $\{\boldsymbol{x}_{i1}, \boldsymbol{x}_{i2}, \ldots, \boldsymbol{x}_{iK}\}$, where $\boldsymbol{x}_{i1} \in \boldsymbol{X}_1$, $\boldsymbol{x}_{i2} \in \boldsymbol{X}_2, \ldots$, and $\boldsymbol{x}_{iK} \in \boldsymbol{X}_K$.

- *Definition 1(Factorial Cost)*: For a given task T_j, the *factorial cost* Ψ_{ij} of individual p_i is given by $\Psi_{ij} = \lambda \cdot \delta_{ij} + F_{ij}$; where λ is a large penalizing multiplier, F_{ij} and δ_{ij} are the objective value and the total constraint violation, respectively, of p_i with respect to T_j. Accordingly, if p_i is feasible with respect to T_j (zero constraint violation), we have $\Psi_{ij} = F_{ij}$.
- *Definition 2(Factorial Rank)*: The *factorial rank* r_{ij} of p_i on task T_j is simply the index of p_i in the list of population members sorted in ascending order with respect to factorial cost Ψ_{ij}.

Note that, while assigning factorial ranks, whenever $\Psi_{1j} = \Psi_{2j}$ for a pair of individuals p_1 and p_2, the parity is resolved by random tie-breaking.

- *Definition 3(Skill Factor)*: The *skill factor* τ_i of p_i is the one task, amongst all other tasks in a K-factorial environment, with which the individual is associated. If p_i is evaluated for all K tasks then $\tau_i = argmin_j\{r_{ij}\}$, where $j \in \{1, 2, \ldots, K\}$.
- *Definition 4(Scalar Fitness)*: The *scalar fitness* of p_i in a multitasking environment is given by $\varphi_i = 1/r_{iT}$, where $T = \tau_i$. Notice that $max\{\varphi_i\} = 1$.

Once the fitness of every individual has been scalarized according to Definition 4, performance comparison can then be carried out in a straightforward manner. For example, individual p_1 will be considered to dominate individual p_2 in multifactorial sense simply if $\varphi_1 > \varphi_2$.

It is important to note that the procedure described heretofore for comparing individuals is not absolute. As the factorial rank of an individual, and implicitly its scalar fitness, depends on the performance of every other individual in the population, the comparison is in fact population dependent. Nevertheless, the procedure guarantees that if an individual p^* uniquely attains the global optimum of any task then $\varphi^* = 1$, which implies that $\varphi^* \geq \varphi_i$ for all $i \in \{1, 2, \ldots, |P|\}$. Therefore, it can be said that the proposed technique is indeed consistent with the ensuing definition of multifactorial optimality.

- *Definition 5(Multifactorial Optimality)*: An individual p^* is considered to be optimum in multifactorial sense if there exists at least one task in the K-factorial environment which it globally optimizes.

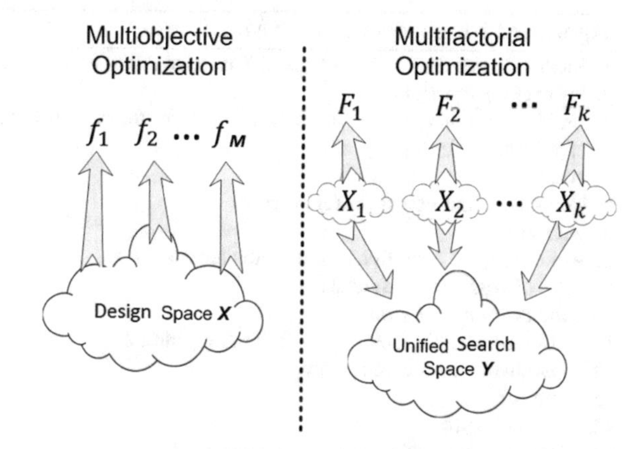

Fig. 1 Multi-objective optimization typically comprises a single design space encompassing all objective functions. On the other hand, multitask optimization unifies (into $\boldsymbol{Y}$) multiple heterogeneous design spaces belonging to distinct tasks [13]

2.1 Multitask Versus Multi-objective Optimization

Since multitask and multi-objective optimization are both concerned with processing a set of objective functions, a conceptual overlap may be seen to exist between them. However, it must be observed that there exists a vital difference between the fundamental principles of the two paradigms. While MFO aims to *leverage upon the implicit parallelism of population-based search to exploit the underlying commonalities and/or complementarities between multiple separate (but possibly similar) optimization tasks*, the formulation of a multi-objective optimization problem and its associated solution algorithms (such as any MOEA) attempt to effectively resolve conflicts among competing objectives of *the same task*. An illustration summarizing the statement is depicted in Fig. 1. The key ingredient distinguishing the two paradigms is the simultaneous existence of multiple heterogeneous design spaces in the case of multitasking, each corresponding to a distinct task. On the other hand, for the case of multi-objective optimization, there typically exists a single design space for a given task of interest, with all objective functions depending on variables contained within that space. Furthermore, note that a multitasking environment could potentially include a multi-objective optimization task as one among many other concurrent tasks, which highlights the greater generality of the proposed paradigm.

3 Multifactorial Evolution: A Framework for Effective Multitasking

In this section we describe the Multifactorial Evolutionary Algorithm (MFEA), an effective multitasking framework that draws upon the bio-cultural models of multifactorial inheritance [15, 16]. As the workings of the approach are based on the transmission of biological as well as cultural building blocks from parents to their

Algorithm 1 Pseudocode of the MFEA

1: Randomly generate n individuals in $\boldsymbol{Y}$ to form initial population P_0
2: **for every** p_j in P_0 **do**
3: Assign skill factor $\tau_j = \mod(j, K) + 1$, for the case of K tasks
4: Evaluate p_j for task τ_j only
5: **end for**
6: Compute scalar fitness φ_j for every p_j
7: Set $t = 0$
8: **while** stopping conditions are not satisfied **do**
9: $C_t = \text{Crossover} + \text{Mutate}(P_t)$
10: **for every** c_j in C_t **do**
11: Determine skill factor $\tau_j \rightarrow$ Refer Algorithm 2
12: Evaluate c_j for task τ_j only
13: **end for**
14: $R_t = C_t \cup P_t$
15: Update scalar fitness of all individuals in R_t
16: Select N fittest members from R_t to form P_{t+1}
17: Set $t = t + 1$
18: **end while**

offspring, the MFEA is regarded as belonging to the realm of *memetic computation* [19, 20]—a field that has recently emerged as a successful computational paradigm synthesizing Darwinian principles of natural selection with the notion of memes, as put forth by Richard Dawkins, as the basic unit of cultural evolution [21]. An overview of the procedure is provided next.

As shown in Algorithm 1, the MFEA starts by randomly creating a population of n individuals in the unified search space $\boldsymbol{Y}$. Moreover, each individual in the initial population is pre-assigned a specific skill factor (see Definition 3) in a manner that guarantees every task to have uniform number of representatives. We would like to emphasize that the skill factor of an individual (i.e., the task with which the individual is associated) is viewed as a computational representation of its pre-assigned cultural trait. The significance of this step is to ensure that an individual is only evaluated with respect to a single task (i.e., only its skill factor) amongst all other tasks in the multitasking environment. Doing so is considered practical since evaluating every individual exhaustively for every task will generally be computationally demanding, especially when K (the number of tasks in the multitasking environment) becomes large. The remainder of the MFEA proceeds similarly to any standard evolutionary procedure. In fact, it must be mentioned here that the underlying genetic mechanisms may be borrowed from any of the plethora of population-based algorithms available in the literature, keeping in mind the properties and requirements of the multitasking problem at hand. The only significant deviation from a traditional approach occurs in terms of offspring evaluation which accounts for cultural traits via individual skill factors.

3.1 Offspring Evaluation in the MFEA

Following the memetic phenomenon of *vertical cultural transmission* [17–19], offspring in the MFEA experience strong cultural influences from their parents, in addition to inheriting their genes. In gene-culture co-evolutionary theory, vertical cultural transmission is viewed as a mode of inheritance that operates in tandem with genetics, and leads to the phenotype of an offspring being directly influenced by the phenotype of its parents. The algorithmic realization of the aforementioned notion is achieved in the MFEA via a *selective imitation strategy*. In particular, selective imitation is used to mimic the commonly observed phenomenon that offspring tend to imitate the cultural traits (i.e., skill factors) of their parents. Accordingly, in the MFEA, an offspring is only decoded (from the unified genotype space $\boldsymbol{Y}$ to a task-specific phenotype space) and evaluated with respect to a single task with which at least one of its parents is associated. As has been mentioned earlier, selective evaluation plays a role in managing the computation expense of the MFEA. A summary of the steps involved is provided in Algorithm 2.

Algorithm 2 Vertical cultural transmission via selective imitation

Consider offspring $c \in C_t$ where $c = \text{Crossover} + \text{Mutate}(p_1, p_2)$
1: Generate a random number $rand$ between 0 and 1
2: **if** $rand \leq 0.5$ **then**
  c imitates skill factor of p_1
3: **else**
  c imitates skill factor of p_2
4: **end if**

3.2 Search Space Unification and Cross-Domain Decoding Exemplars

The core motivation behind the evolutionary multitasking paradigm is the autonomous exploitation of known or latent commonalities and/or complementarities between distinct (but possibly similar) optimization tasks for achieving faster and better convergence characteristics. One of the possible means of harnessing the available synergy, at least from an evolutionary perspective, is through implicit genetic transfer during crossover operations. However, for the relevant knowledge to be transferred across appropriately, i.e., to ensure effective multitasking, it is pivotal to first describe a genotypic unification scheme that suits the requirements of the multitasking problem at hand. In particular, the unification serves as a higher-level abstraction that constitutes a *meme space*, wherein building blocks of encoded knowledge are processed and shared across different optimization tasks. This perspective is much in alignment with the workings of the human brain, where knowledge pertaining to different tasks are abstracted, stored, and re-used for relevant problem solving exercises whenever needed.

Unification implies that genetic building blocks [22] corresponding to different tasks are contained within a single pool of genetic material, thereby facilitating the MFEA to process them in parallel. To this end, assuming the search space dimensionality of the j-th optimization task (in isolation) to be D_j, a unified search space $\boldsymbol{Y}$ comprising K (traditionally distinct) tasks may be defined such that $D_{multitask} = max_j\{D_j\}$, where $j \in \{1, 2, \ldots, K\}$. In other words, while handling K optimization tasks simultaneously, the chromosome $\boldsymbol{y} \in \boldsymbol{Y}$ of an individual in the MFEA is represented by a vector of $D_{multitask}$ variables. While addressing the j-th task, we simply extract D_j variables from the chromosome and decode them into a meaningful solution representation for the underlying optimization task. In most cases, an appropriate selection of D_j task-specific variables from the list of $D_{multitask}$ variables is crucial for the success of multitasking. For instance, if two distinct variables belonging to two different tasks have similar phenotypic meaning, then they should intuitively be associated to the same variable in the unified search space $\boldsymbol{Y}$. On the other hand, in many naive cases where no a priori understanding about the phenotype space is available, simply extracting the *first* D_j variables from the chromosome can oftentimes be a viable alternative [12].

In what follows, we demonstrate how chromosomes in a unified genotype space can be decoded into meaningful task-specific solution representations when a *random-key unification scheme* [23] is adopted. According to the random-key scheme, each variable of a chromosome is simply encoded by a *continuous value* in the range [0, 1]. The salient feature of this representation is that it elegantly accommodates a wide variety of problems in continuous as well as discrete optimization, thereby laying the foundation for a cross-domain multitasking platform. Some decoding examples for continuous and popular instantiations of combinatorial optimization shall be discussed hereafter. At this juncture, it must however be emphasized that the concept of multitasking is not necessarily tied to cross-domain optimization. In fact, domain-specific schemes can indeed be used (often with greater success) when all constitutive tasks belong to similar domains.

3.2.1 Decoding for Continuous Optimization Problems

In the case of continuous optimization, decoding can be achieved in a straightforward manner by linearly mapping each random-key from the genotype space to the box-constrained phenotype space of the relevant optimization task [12].

3.2.2 Decoding for Discrete Sequencing Problems

In the domain of combinatorial optimization, sequencing problems include a variety of classical examples such as the Travelling Salesman (TSP), Job-Shop Scheduling (JSP), Quadratic Assignment (QAP), Vehicle Routing (VRP), etc. The common feature of these problems is that they involve the *ordering* of a finite set of distinct entities in a manner that optimizes a given objective function. The applicability of the real parameter random-key chromosome representation scheme to discrete problems of this kind was perhaps first investigated in [23]. In particular, it was observed that under any real-coded variation operation, the decoding procedure ensures feasibility of the generated offspring. This outcome is in contrast to domain-specific representations of sequencing problems wherein specially designed variation operators are needed to ensure offspring feasibility. As a consequence, the random-key representation has found notable interest over the past two decades in the field of operations research [24–26].

For an illustration of the decoding scheme, consider a case where 5 distinct entities are to be ordered optimally. To this end, a sample random-key chromosome in the MFEA may look like $\mathbf{y} = (0.1, 0.7, 0.2, 0.9, 0.04)$, such that the first entity is labeled as 0.1, the second entity is labeled as 0.7, the third is labeled as 0.2, and so on. Following the technique suggested in [23], the order of entities encoded by the chromosome $\mathbf{y}$ is given by the sequence $\boldsymbol{s} = (5, 1, 3, 2, 4)$. In other words, the sequence can be deduced simply by *sorting* the random-key labels in ascending order. Each entity is assigned an index in $\boldsymbol{s}$ that corresponds to the position of its label in the sorted list.

3.3 Implicit Knowledge Transfer in the MFEA

For any proposed unification scheme to be useful for multitasking, a matter of critical importance is the means of knowledge transfer in the unified space. In this regard, it has been stated that knowledge transfer across two or more optimization tasks, being simultaneously solved in the MFEA, occurs in the form of implicit genetic exchange between cross-cultural parents undergoing crossover [13]. While there are a plethora of such operators available in the literature, many of which exploit unique features of the underlying optimization tasks, herein we focus on the mechanics of the well-established simulated binary crossover (SBX) operator [27] from the standpoint of multitasking.

A salient feature of the SBX operator is that it emphasizes (with high probability) on creating offspring that are located close to their parents [28]. In other words, in a continuous search space, it is often the case that a generated offspring possesses genetic material that is in close proximity to at least one of its parents. With this background, consider the situation in Fig. 2 where two parents p_1 and p_2, with different cultural traits or skill factors (recall Definition 3), undergo crossover in a hypothetical 2-D unified search space. In particular, p_1 is assigned skill factor τ_1 while p_2 is assigned skill factor τ_2, with $\tau_1 \neq \tau_2$. Further, a pair of offspring, namely c_1 and c_2, is generated in the neighborhood of the parents by the SBX operator. Notice that c_1 is found to inherit much of its genetic material from p_1, while c_2 is found to inherit much of its genetic material from p_2. In such a scenario, if c_1 imitates the skill factor of p_2 (i.e., if c_1 is evaluated for τ_2) and/or if c_2 imitates the skill factor of p_1 (i.e., if c_2

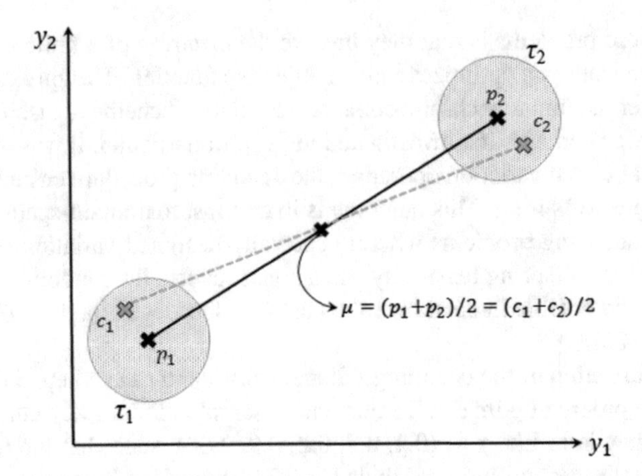

Fig. 2 Parent candidates p_1 and p_2 undergo standard SBX crossover to produce offspring c_1 and c_2 that are located close to their parents with high probability. Parent p_1 possesses skill factor τ_1 and p_2 possesses skill factor τ_2 with $\tau_1 \neq \tau_2$, thereby creating a multicultural environment for offspring to be reared in. Now, if c_1 imitates p_1 and/or if c_2 imitates p_2, then implicit genetic transfer is said to occur between the two tasks [13]

is evaluated for τ_1), then implicit transfer of knowledge occurs between the two tasks. At this juncture, if the genetic material corresponding to τ_1 (carried by c_1) is found to be useful for τ_2, or vice versa, then the transfer is deemed beneficial. Thereafter, the evolutionary selection pressure takes over to ensure that the positively transferred knowledge survives through generations. On the other hand, if the transfer turns out to be unproductive, the fundamental property of evolution is to eliminate the weak (negatively transferred [29–31]) genes by the natural process of survival of the fittest.

3.4 A Summary of the Salient Features of Evolutionary Multitasking

Standard EAs typically generate a large population of candidate solutions, all of which are unlikely to be competent for the task at hand. In contrast, in a multitasking environment, wherein all constitutive tasks are assimilated into a unified search space, it is intuitively more probable that a randomly generated or genetically modified individual is competent for at least one task. The mechanisms of the MFEA leverage upon this observation by effectively coordinating the search via the metaphorical interactions of genetic and cultural factors, thereby facilitating enhanced productivity in decision making processes in real-world settings.

Interestingly, during the combined optimization process it may so happen that the refined genetic material created within individuals of a particular skill factor (i.e., of a particular cultural trait) may also be useful for another group of individuals with a different skill factor. Thus, in such situations, the scope for implicit genetic transfer across tasks can potentially lead to accelerated convergence characteristics and/or the discovery of hard to find global optima. For the MFEA in particular, the transfer of genetic material occurs whenever cross-cultural parents with different skill factors undergo chromosomal crossover, as described in the previous subsection.

Practical scenarios amenable to multitasking are likely to occur in a variety of domains, including engineering, business, operations, etc., wherein optimization tasks with essentially identical underlying characteristics recur in large numbers. As per traditional practices, the knowledge contained in these related tasks is generally ignored by taking a tabula rasa approach to optimization. To this end, evolutionary multitasking provides a novel means of harnessing the so-far untapped source of knowledge, thereby opening doors to a plethora of real-world opportunities, some of which shall be showcased next.

4 Scope for Multitasking in the Real-World

Humans demonstrate cognitive multitasking capabilities on a daily basis. In [12], this anthropic phenomenon was realized computationally in the form of evolutionary multitasking for optimization. In order to emphasize the considerable real-world scope of multitasking, we present some guiding thoughts to aid effective utilization of the concept. It is contended that insights for a variety of practical applications can naturally be inferred from our discussions.

Without loss of generality, consider a hypothetical 2-factorial scenario where the first task is labeled $\mathcal{T}_1$ and the second task is labeled as $\mathcal{T}_2$. The setup of the multitasking environment is depicted in Fig. 3. Therein, notice the presence of a unified genotype space $\boldsymbol{Y}$ that encodes solutions to each of the constitutive tasks. In particular, $\boldsymbol{x}_1$ represents a solution in the phenotype space of $\mathcal{T}_1$ while $\boldsymbol{x}_2$ represents a solution in the phenotype space of $\mathcal{T}_2$. With this background, we categorize multitasking problem instances based on the amount of overlap in the phenotype space. We quantify the *overlap* (χ) as the number of variables in a task-specific solution space that have similar phenotypic meaning with respect to the other task, i.e., $\chi = |\boldsymbol{x}_{overlap}|$, leading to three broad categories, namely, complete, partial, and no overlap.

4.1 Complete Overlap in Phenotype Space

The first scenario we consider is perhaps the most intuitively pleasing application domain for evolutionary multitasking. In particular, we assume $\boldsymbol{x}_1 \backslash \boldsymbol{x}_{overlap} = \boldsymbol{x}_2 \backslash \boldsymbol{x}_{overlap} = \varnothing$ in Fig. 3. Accordingly, the only feature distinguishing the tasks is the set of task-specific *auxiliary variables* which are not explicitly part of the search space but describe the background in which the optimization tasks play out. A variety of possible real-world manifestations of this category in fields such as complex

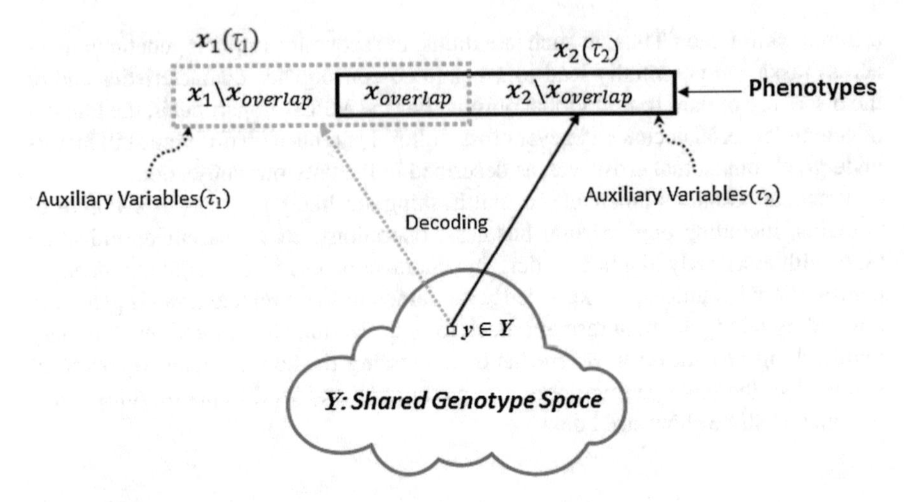

Fig. 3 Setup of a 2-factorial environment. The overlap in phenotype space represents the variables that have similar phenotypic interpretation with respect to either task. Note that although the overlapping variables need not bear identical numeric values for both tasks, they often provide the scope for useful genetic transfer due to similarities in their underlying behavior

engineering design and operations research have been discussed in [13]. In the present chapter, we delve into recent advancements in other areas of interest that have not been reviewed in previous papers.

A promising approach for improving optimization performance is the creation of *artificial helper (or catalyst) tasks that can aid the search process for a target optimization task of interest*, i.e., when both are combined in a single multitasking environment. While this possibility has been exploited in the field of machine learning [32], little has been done in the context of optimization. The lack of related approaches in optimization is particularly surprising given the availability of population-based methods that are endowed with the power of implicit parallelism. In light of this fact, preliminary investigations show that combining a target single-objective optimization task together with an artificially created multi-objective reformulation of the same task can improve convergence characteristics [33]. A representative example is depicted in Fig. 4 for a TSP instance where the target task and the helper task have completely overlapping phenotype spaces. In essence, the multi-objective reformulation, which has often been found to remove local optima [34], aids performance by leveraging on the scope for implicit genetic transfer.

In addition to the above, a recent study in bi-level optimization has shown the potential utility of evolutionary multitasking therein [14]. It was found that the notion of multitasking naturally emerges in the realm of evolutionary bi-level optimization where several lower level optimization tasks are to be solved with respect to different upper level population members. In particular, lower level tasks corresponding to neighboring upper level individuals, such as those belonging to the same cluster (as shown in Fig. 5), are likely to possess useful underlying commonalities that can

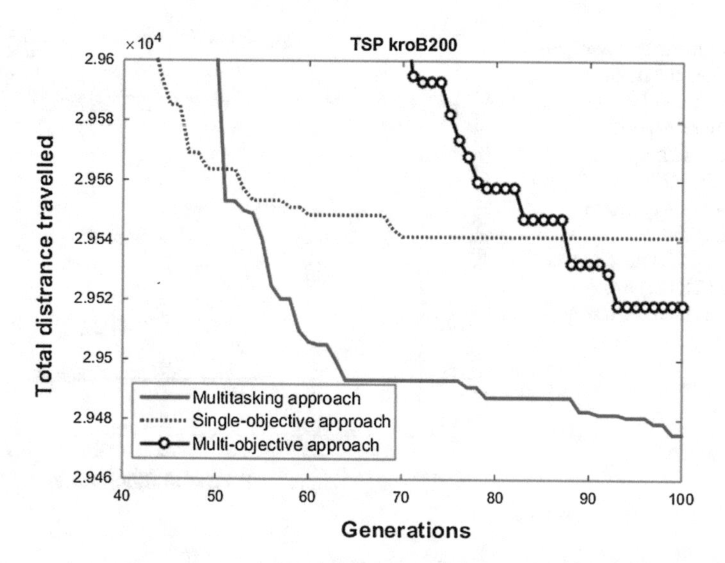

Fig. 4 Convergence trends for single-objective, multi-objective, and multitasking approaches for TSP kroB200. Multitasking harnesses the unique advantages of the single-objective and multi-objective formulations to accelerate convergence. Here, the artificially formulated multi-objective task acts as a catalyst during multitasking

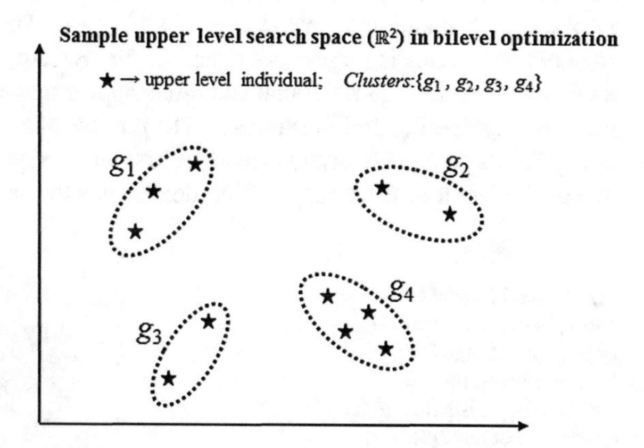

Fig. 5 In evolutionary bilevel optimization, lower level tasks corresponding to closely located upper level individuals (such as those belonging to the same cluster) are likely to possess commonalities that are exploitable by multitasking

be exploited via multitasking. The efficacy of the proposition was demonstrated by a proof-of-concept case study from the composites manufacturing industry which led to a computational cost saving of nearly 65 % for an expensive simulation-based optimization exercise [14]. A representative plot comparing the convergence trends achieved in practical bi-level optimization with and without evolutionary multitasking is provided in Fig. 6.

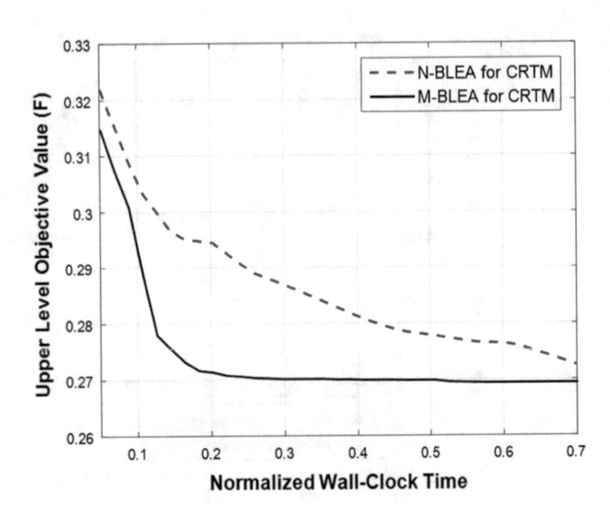

Fig. 6 Comparing averaged convergence trends of a standard Nested-Bilevel Evolutionary Algorithm (N-BLEA) with a Multitasking-Bilevel Evolutionary Algorithm (M-BLEA) for a Compression Resin Transfer Molding (CRTM) based composites manufacturing cycle [14]

4.2 Partial Overlap in Phenotype Space

Next, we consider the case where the phenotype spaces of constitutive tasks are only partially overlapping. For the 2-factorial setup in Fig. 3, this implies that $\boldsymbol{x}_1 \backslash \boldsymbol{x}_{overlap} \neq \varnothing$ and/or $\boldsymbol{x}_2 \backslash \boldsymbol{x}_{overlap} \neq \varnothing$ and $\chi \geq 1$. Thus, the transferrable knowledge between tasks is largely contained in the overlapping region, i.e., in $\boldsymbol{x}_{overlap}$. Real-world instantiations of such situations appear aplenty in the *conceptualization phase* of engineering design exercises. The process of conceptualization, as depicted in Fig. 7, is a human creativity driven preliminary design stage dealing with the formulation of an idea or concept which determines the scope of a project in terms of

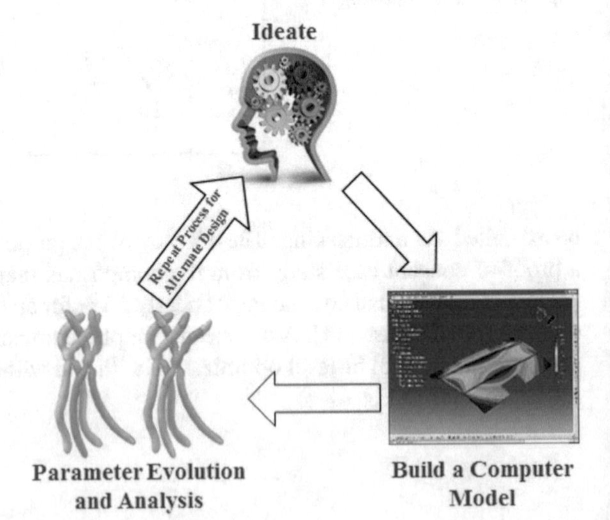

Fig. 7 Workflow of the conceptualization phase in engineering design [13]. Immense scope for multitasking exists due to the emergence of multiple alternative concepts to be analyzed. The concepts are likely to share some underlying commonalities as they all cater to the same product or process. This knowledge may be harnessed during multitasking to accelerate the design process

desired design features and requirements [35–37]. Typically, numerous alternative approaches will be proposed and analyzed before agreeing upon the single most suitable one. In these situations, the scope for evolving similar concepts via multitasking is quite intuitive, especially because several overlapping (i.e., recurring) design variables appear in different conceptual designs. Therefore, useful transferrable knowledge is instinctively known to exist among the tasks as they pertain to the same underlying product or process [13].

4.3 No Overlap in Phenotype Space (Blind Multitasking)

In both categories discussed so far, it is generally possible to make an a priori inference about the existence of transferrable knowledge that can be exploited by the process of multitasking. However, in many other real-world applications, it may be extremely difficult, if not impossible, to make such prior judgment about the complementarity between different optimization tasks. Multitasking instances belonging to the third category of no overlap in phenotype space, i.e., $\boldsymbol{x}_{overlap} = \emptyset$, are examples of such blind multitasking. However, even in these cases, it is noted that some latent complementarity between tasks may continue to exist in the unified genotype space. Thus, it often makes sense to allow evolution to take over and autonomously harness the complementarities whenever available, without the need to explicitly identify and inject domain knowledge into the algorithm. Needless to say, the execution of blind multitasking in the proposed naïve manner raises the fear of predominantly negative transfer. Whether the potential for enhanced productivity is sufficient to subdue such fears remains to be seen in the future. In the long run however, an *ideal* evolutionary multitasking engine is envisaged to be a complex adaptive system that is capable of inferring and appropriately responding to inter-task relationships on the fly, with its overall performance being at least comparable to the single-task solvers of the present day.

For the purpose of demonstration, we present a multitasking instance where performance enhancements are achieved despite the lack of any apparent overlap in the phenotype spaces of constitutive tasks. The example combines a pair of combinatorial optimization problems. As is well known, combinatorial problems possess complex objective function landscapes that are generally difficult to analyze. Thus, in most cases it is extremely challenging to make any prior inference about the availability of transferrable knowledge across tasks. Nevertheless, it can be concluded from the convergence trends in Fig. 8 that even in such cases of blind multitasking performance enhancement is achievable via the MFEA.

The 2-factorial problem depicted in Fig. 8 comprises a TSP (kroA200) and JSP (la39). For both tasks, the single-tasking approach is found to consistently get trapped in a local optimum. On the other hand, the diversified search facilitated by multitasking substantially improves performance characteristics, primarily as a result of the constant transfer of genetic material from one task to the other. It is therefore contended that while no decipherable complementarity exists between the tasks when

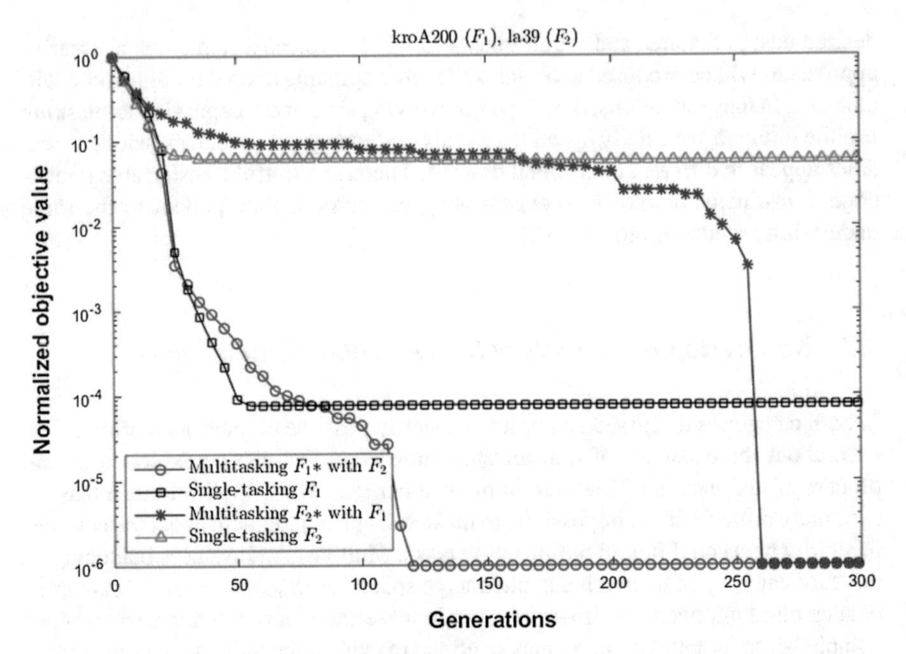

Fig. 8 Averaged convergence trends achieved while single-tasking and while multitasking across combinatorial optimization problems occurring in complex supply chain networks: TSP (kroA200) and JSP (la39) [13]

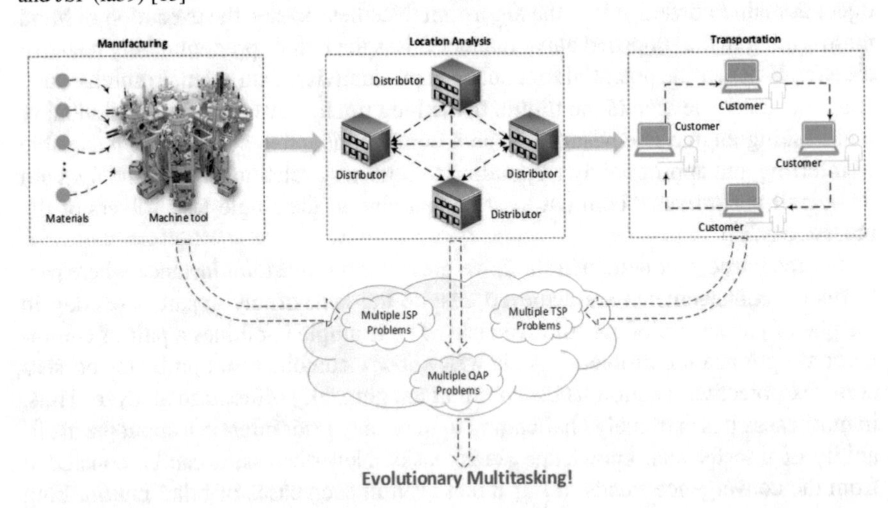

Fig. 9 Complex multi-echelon supply chain networks provide promising future prospects for the application of evolutionary multitasking [13]

viewed in the phenotype space, some latent complementarity may emerge in the unified genotype space. A real-world setting where the need to multitask across such seemingly disparate problems may arise is that of complex multi-echelon supply chain networks. The increase in productivity can help ease bottlenecks in decision making across multiple silos at once. Accordingly, as illustrated in Fig. 9, the domain of supply chain management can be a notable future beneficiary of evolutionary multitasking. For instance, while a TSP may represent a transportation (or logistics) silo of a supply chain, the JSP may represent a manufacturing silo, together forming key ingredients of the overall network.

5 Conclusions and Directions for Future Research

Evolutionary multitasking is a novel optimization paradigm that, albeit in its in-fancy, is showcasing significant promise with regard to unleashing the true power of implicit parallelism of population-based search [38]. To highlight the fact that each task in a multitasking environment presents an additional factor influencing the evolution of single population of individuals, the paradigm has also been formally labeled as Multifactorial Optimization (MFO). Sharing similar motivations as the field of multitask learning, MFO provides the scope for exploiting the underlying commonalities and/or complementarities between different (but possibly similar) optimization tasks, thereby achieving accelerated convergence characteristics in comparison to standard single-task optimizers. Furthermore, the quality of results obtained in a variety of domains of practical interest strongly encourages more comprehensive research pursuits in the future. It is envisaged that with increasing contributions from the community of EC researchers, as well as from the computer science and engineering communities at large, the notion of multitasking has the potential to change the current landscape of optimization techniques by seamlessly incorporating the scope of autonomous knowledge adaptation from various sources. In particular, it is contended that an artificial (computational) multitasking engine may be capable of retaining many of the advantages of cognitive multitasking, while effectively overcoming its potential perils.

In summary, it is recognized that so far we have merely scratched the surface of a potentially rich research topic. Rigorous examination of several practical and theoretical aspects of the paradigm is needed in the future. To begin with, a fundamental question that may arise in the mind of a practitioner is whether multitasking will always improve performance. In this regard, it must be noted that evolutionary multitasking acts as a means of harnessing the inductive bias provided by other optimization tasks in the same multitasking environment. Thus, while some inductive biases are helpful, some other inductive biases may hurt [10]. In fact, in the current simplistic description of the MFEA, we have indeed encountered some counter examples where the observed performance deteriorates during multitasking. However, in the long run, an ideal evolutionary multitasking engine is conceived to be an adaptive system that will be capable of estimating and autonomously responding to the level of complementarity between tasks on the fly. Thus, with the aim of enhancing productivity in complex decision making environments, it is the design of such intelligent algorithms that shall form the crux of our future research endeavours.

References

1. Dzubak, C.M., et al.: Multitasking: the good, the bad, and the unknown. J. Assoc. Tutoring Prof. **1**(2), 1–12 (2008)
2. Just, M.A., Buchweitz, A.: What brain imaging reveals about the nature of multitasking (2011)
3. Bäck, T., Hammel, U., Schwefel, H.-P.: Evolutionary computation: comments on the history and current state. IEEE Trans. Evol. Comput. **1**(1), 3–17 (1997)
4. Goldberg, D.E., et al.: Genetic Algorithms in Search Optimization and Machine Learning, vol. 412. Addison-Wesley, Reading (1989)
5. Srinivas, M., Patnaik, L.M.: Genetic algorithms: a survey. Computer **27**(6), 17–26 (1994)
6. Bertoni, A., Dorigo, M.: Implicit parallelism in genetic algorithms. Artif. Intell. **61**(2), 307–314 (1993)
7. Deb, K.: Multi-objective Optimization Using Evolutionary Algorithms, vol. 16. Wiley, New York (2001)
8. Branke, J.: MCDA and multiobjective evolutionary algorithms. Multiple Criteria Decision Analysis, pp. 977–1008. Springer, New York (2016)
9. Li, B., Li, J., Tang, K., Yao, X.: Many-objective evolutionary algorithms: a survey. ACM Comput. Surv. (CSUR) **48**(1), 13 (2015)
10. Caruana, R.: Multitask learning. Mach. Learn. **28**(1), 41–75 (1997)
11. Ben-David, S., Schuller, R.: Exploiting task relatedness for multiple task learning. Learning Theory and Kernel Machines, pp. 567–580. Springer, New York (2003)
12. Gupta, A., Ong, Y.-S., Feng, L.: Multifactorial evolution: towards evolutionary multitasking. Accepted IEEE Trans. Evol. Comput. **10**, 1109 (2015)
13. Ong, Y.-S., Gupta, A.: Evolutionary multitasking: a computer science view of cognitive multitasking. Cogn. Comput. **8**(2), 125–142 (2016)
14. Gupta, A., Mańdziuk, J., Ong, Y.-S.: Evolutionary multitasking in bi-level optimization. Complex Intell. Syst. **1**(1–4), 83–95 (2015)
15. Rice, J., Cloninger, C., Reich, T.: Multifactorial inheritance with cultural transmission and assortative mating. i. description and basic properties of the unitary models. Am. J. Hum. Genet. **30**(6), 618 (1978)
16. Cloninger, C.R., Rice, J., Reich, T.: Multifactorial inheritance with cultural transmission and assortative mating. ii. a general model of combined polygenic and cultural inheritance. Am. J. Hum. Genet. **31**(2), 176 (1979)
17. Cavalli-Sforza, L.L., Feldman, M.W.: Cultural versus biological inheritance: phenotypic transmission from parents to children (a theory of the effect of parental phenotypes on children's phenotypes). Am. J. Hum. Genet. **25**(6), 618 (1973)
18. Feldman, M.W., Laland, K.N.: Gene-culture coevolutionary theory. Trends Ecol. Evol. **11**(11), 453–457 (1996)
19. Chen, X., Ong, Y.-S., Lim, M.-H., Tan, K.C.: A multi-facet survey on memetic computation. IEEE Trans. Evol. Comput. **15**(5), 591–607 (2011)
20. Ong, Y.-S., Lim, M.H., Chen, X.: Research frontier-memetic computation past, present and future. IEEE Comput. Intell. Mag. **5**(2), 24 (2010)
21. Dawkin, R.: The Selfish Gene, vol. 1, p. 976. Oxford University Press, Oxford (1976)
22. Iqbal, M., Browne, W.N., Zhang, M.: Reusing building blocks of extracted knowledge to solve complex, large-scale boolean problems. IEEE Trans. Evol. Comput. **18**(4), 465–480 (2014)
23. Bean, J.C.: Genetic algorithms and random keys for sequencing and optimization. ORSA J. Comput. **6**(2), 154–160 (1994)

24. Gonçalves, J.F., Resende, M.G.: Biased random-key genetic algorithms for combinatorial optimization. J. Heuristics **17**(5), 487–525 (2011)
25. Snyder, L.V., Daskin, M.S.: A random-key genetic algorithm for the generalized traveling salesman problem. Eur. J. Oper. Res. **174**(1), 38–53 (2006)
26. Gonçalves, J.F., Resende, M.G.: A parallel multi-population biased random-key genetic algorithm for a container loading problem. Comput. Oper. Res. **39**(2), 179–190 (2012)
27. Deb, K., Agrawal, R.B.: Simulated binary crossover for continuous search space. Complex Syst. **9**(3), 1–15 (1994)
28. Deb, K., Sindhya, K., Okabe, T.: Self-adaptive simulated binary crossover for real-parameter optimization. In: Proceedings of the 9th Annual Conference on Genetic and Evolutionary Computation, pp. 1187–1194. ACM (2007)
29. Pan, S.J., Yang, Q.: A survey on transfer learning. IEEE Trans. Knowl. Data Eng. **22**(10), 1345–1359 (2010)
30. Feng, L., Ong, Y.-S., Lim, M.-H., Tsang, I.W.: Memetic search with interdomain learning: a realization between CVRP and CARP. IEEE Trans. Evol. Comput. **19**(5), 644–658 (2015)
31. Feng, L., Ong, Y.-S., Tan, A.-H., Tsang, I.W.: Memes as building blocks: a case study on evolutionary optimization+transfer learning for routing problems. Memet. Comput. **7**(3), 159–180 (2015)
32. Krawiec, K., Wieloch, B.: Automatic generation and exploitation of related problems in genetic programming. In: IEEE Congress on Evolutionary Computation (CEC), pp. 1–8. IEEE (2010)
33. Da, B., Gupta, A., Ong, Y.-S., Feng, L.: Evolutionary multitasking across single and multi-objective formulations for improved problem solving. In: 2010 IEEE Congress on Evolutionary Computation (CEC). IEEE (2016)
34. Knowles, J.D., Watson, R.A., Corne, D.W.: Reducing local optima in single-objective problems by multi-objectivization. Evolutionary Multi-criterion Optimization, pp. 269–283. Springer, New York (2001)
35. Avigad, G., Moshaiov, A.: Set-based concept selection in multi-objective problems: optimality versus variability approach. J. Eng. Des. **20**(3), 217–242 (2009)
36. Avigad, G., Moshaiov, A., Brauner, N.: MOEA-based approach to delayed decisions for robust conceptual design. Applications of Evolutionary Computing, pp. 584–589. Springer, New York (2005)
37. Avigad, G., Moshaiov, A.: Interactive evolutionary multiobjective search and optimization of set-based concepts. IEEE Trans. Syst. Man Cybern. Part B: Cybern. **39**(4), 1013–1027 (2009)
38. Gupta, A., Ong, Y.-S., Feng, L., Tan, K.C.: Multiobjective multifactorial optimization in evolutionary multitasking (2016)

Practical Applications in Constrained Evolutionary Multi-objective Optimization

Arun Kumar Sharma, Rituparna Datta, Maha Elarbi, Bishakh Bhattacharya and Slim Bechikh

Abstract Constrained optimization is applicable to most real world engineering science problems. An efficient constraint handling method must be robust, reliable and computationally efficient. However, the performance of constraint handling mechanism deteriorates with the increase of multi-modality, non-linearity and non-convexity of the constraint functions. Most of the classical mathematics based optimization techniques fails to tackle these issues. Hence, researchers round the globe are putting hard effort to deal with multi-modality, non-linearity and non-convexity, as their presence in the real world problems are unavoidable. Initially, Evolutionary Algorithms (EAs) were developed for unconstrained optimization but engineering problems are always with certain type of constraints. The in-dependability of EAs to the structure of problem has led the researchers to re-think in applying the same to the problems incorporating the constraints. The constraint handling techniques have been successfully used to solve many single objective problems but there has been

A.K. Sharma (✉) · B. Bhattacharya
Smart Materials Structure and Systems (SMSS) Laboratory, Department of Mechanical Engineering, Indian Institute of Technology Kanpur, Kanpur 208016, India
e-mail: sarun@iitk.ac.in; arunme65@gmail.com

B. Bhattacharya
e-mail: bishakh@iitk.ac.in

R. Datta
Graduate School of Knowledge Service Engineering, Department of Industrial and Systems Engineering, Korean Advanced Institute of Science and Technology (KAIST), 291 Daehak-ro, Yuseong-gu, Daejeon 34141, Republic of Korea
e-mail: rdatta@kaist.ac.kr; rdatta@iitk.ac.in

R. Datta
Department of Mechanical Engineering, Indian Institute of Technology Kanpur, Kanpur, Uttar Pradesh, India

M. Elarbi · S. Bechikh
SOIE Lab, Computer Science Department, ISG-Tunis, University of Tunis, Bouchoucha City, 2000 Le Bardo, Tunis, Tunisia
e-mail: arbi.maha@yahoo.com

S. Bechikh
e-mail: slim.bechikh@gmail.com

S. Bechikh et al. (eds.), *Recent Advances in Evolutionary Multi-objective Optimization*, Adaptation, Learning, and Optimization 20,
DOI 10.1007/978-3-319-42978-6_6

limited work in applying them to the multi-objective optimization problem. Since for most engineering science problems conflicting multi-objectives have to be satisfied simultaneously, multi-objective constraint handling should be one of the most active research area in engineering optimization. Hence, in this chapter authors have concentrated in explaining the constrained multi-objective optimization problem along with their applications.

Keywords Multi-objective optimization · Constraint handling · Evolutionary algorithms · Practical applications

1 Introduction

Evolutionary Multi-objective Optimization (EMO) has been successfully applied over a wide variety of domains of engineering science in the last two decades [1, 2]. Initially, EAs were developed to solve optimization problems with out the consideration of constraints [3, 4]. However, the role of constraints cannot be denied in real world physical problems which are conflicting in nature in the presence of one or many constraints imposed on it. The constraints crop up in real physical world problems either due to geometric dependency or to fulfill operational requirement. Hence, the researchers have focused their attention to integrate constraint handling techniques in EAs and EMOs. This has led research engineers to exploit these techniques in number of complex problems with non-linear and non-convex functions. The constraint handling techniques incorporated into EMO have been handled with Genetic Algorithms (GAs), Differential Evolution (DE), Particle Swarm Optimization (PSO), Evolutionary Strategy (ES), etc. It is also necessary for the Constrained EMO (C-EMO) to satisfy the properties like feasibility, diversity, convergence and computationally efficiency.

A standard Constrained Multi-objective Optimization Problem (CMOP) can be formulated as follows:

$$\begin{cases} \text{Minimize } f_m(x), & m = 1, \ldots, M, \\ \text{subject to } g_j(x) \geq 0, & j = 1, \ldots, J, \\ \qquad h_k(x) = 0, & k = 1, \ldots, K, \\ \qquad x_i^l \leq x_i \leq x_i^u, & i = 1, \ldots, n. \end{cases} \tag{1}$$

The above nonlinear programming (NLP) equation has n number of variables, m number of objective functions, J number of constraints with inequality type, and K number of equality constraints. The number of objective functions are $f_m(x)$, where as j-th inequality constraint is $g_j(x)$ and $h_k(x)$ is the k-th equality constraint. The variation i-th variable is in the range of $[x_i^l, x_i^u]$. As compared to inequality constraints, equality constraints are the most difficult ones to satisfy using any optimization algorithm as the feasible solution must coincide with the intersection of all equality constraints. As a result, obtaining feasible solution is very critical. The regular way to is to convert equality constraints into appropriate inequality constraint by adding tolerance in the following way:

$$\begin{aligned} h_k(x) &\geq -\epsilon, \\ h_k(x) &\leq \epsilon. \end{aligned} \tag{2}$$

where, ϵ is the user defined tolerance taken as 10^{-3}. Thus, the Eq. (1) is converted to $J + 2K$ number of inequality constraints distributed equally on either side of the original value. Now, the total number of inequality constraints can be written in the following way:

$$\begin{aligned} g_j(x) &\geq 0, \quad j = 1, \ldots, J, \\ g_{J+k}(x) &= |\epsilon_k - h_k(x)| \geq 0, \quad k = 1, \ldots, K. \end{aligned} \tag{3}$$

2 Constraint Handling in Evolutionary Multi-objective Optimization

Among the issues that one can face when handling constrained optimization problems is how to deal with the infeasible solutions that do not satisfy at least one of the constraints. In the specialized literature, a variety of constraint handling techniques have been proposed to deal with this issue [5]. In this section, we propose to review the most common used constraint handling techniques for C-EMO.

2.1 *Penalty Functions*

Penalty functions are one of the most used techniques to solve constrained optimization problems [6]. The main idea behind penalty functions is to favor the selection of feasible solutions by decreasing the fitness of the infeasible solutions in the population. In single objective optimization, a penalty function transforms a constrained optimization problem into an unconstrained one by adding a penalty term to the objective function. The expended objective function to be optimized $F(x)$ can be expressed as follows [5]:

$$F(x) = f(x) + P(x) \tag{4}$$

where the constraint violation measure $P(x)$ can be calculated as follows:

$$P(x) = \sum_{j=1}^{J} r_j \cdot max(0, g_j(x))^2 + \sum_{k=1}^{K} c_k \cdot |h_k(x)| \tag{5}$$

where r_j and c_k are two positive penalty factors.

The principle of penalty functions in multi-objective optimization remains similar to single-objective optimization. However, the penalty factor is added to all the objective functions instead of only one objective. The way of defining the penalty factors differ from a penalty function to another (i.e., static, dynamic, adaptive, co-evolved, fuzzy adapted, etc.). In this subsection, we briefly review some of the most representative works that have employed this method for solving CMOPs [7]. Woldesenbet et al. [8] proposed an adaptive penalty function to solve CMOPs. The proposed penalty function adds a large penalty factor to infeasible solutions if there are a few feasible solutions in the population. Besides, a small penalty factor is added. Moreover, they define new objective function values based on distance measure and the adaptive penalty function. In addition, the authors proposed two other penalty functions for infeasible solutions. The first penalty function is based on the objective functions, while the second one is based on the constraint violation. The performance of the proposed algorithm was tested on seven test problems. The obtained results demonstrated that the algorithm is able to find a set of feasible solutions that are well-spread over the Pareto front. The main advantage of this approach is that there is no parameter tuning in the design of constraint handling. Jan and Zhang [9] introduced a modified version of MOEA/DE [10] called CMOEA/D-DE-ATP, where the replacement and the update scheme are modified in order to deal with CMOPs. CMOEA/D-DE-ATP introduced a penalty function that uses a threshold τ to dynamically control the amount of penalty of infeasible solutions. This penalty function is able to guide the algorithm to search the feasible region and the infeasible region which is near the feasible one. CMOEA/D-DE-ATP has shown its effectiveness on six out of ten constrained test instances in terms of the IGD metric and convergence to the true Pareto front. However, it is proved that the elimination of the threshold affects the performance of the algorithm considerably. Jan et al. [11] proposed a dynamic and adaptive version of the penalty function used in CMOEA/D-DE-ATP which led to a new algorithm called CMOEA/D-DE-TDA. Thus, the threshold value τ is set as the average value of the degree of constraint violations of all infeasible solutions in the neighborhood of a solution. However, CMOEA/D-DE-ATP with the dynamic and adaptive penalty functions fails partly if there are few infeasible solutions in the initial population. Additionally, it fails totally if the whole initial population is highly infeasible due to the presence of hard constraints.

2.2 Superiority of Feasible Solutions

This constraint handling technique was introduced by Powell and Skolnick [12]. It was also employed by Deb [13] to solve single-objective optimization problems. For the multi-objective optimization case it can be expressed as follows:

$$fitness_i(x) = \begin{cases} f_m(x) \quad if \ x \ is \ feasible \\ f^m_{worst} + v(x) \end{cases} \tag{6}$$

where f^m_{worst} represents the m-th objective value of the worst feasible solution in the current population and $v(x)$ is the overall constraint violation. In the case where there is no feasible solutions in the population, f^m_{worst} is set to zero.

In this method, feasible solutions are considered better than the infeasible ones. Hence, the feasible solutions are evolved towards the Pareto optimal front, while the infeasible ones are evolved to towards the feasible region. Deb et al. [14] introduced a constrained domination principle which is based on the superiority of feasible solutions. However, it was demonstrated that the adopted constrained domination can cause a premature convergence. Pal et al. [15] used this constraint handling mechanism to solve a linear antenna array synthesis problem as a MOP. The experimental study has mentioned good results in solving the proposed design problem and in achieving good trade-off solutions.

2.3 *ε-Constraint Method*

The ε-constraint handling technique was introduced by Takahama and Sakai [16]. The basic idea is to transform the constrained optimization problem into an unconstrained optimization problem. It uses an ε parameter for the relaxation of the constraint violations in the earlier stages of evaluation. In fact, infeasible solutions with small overall constraint violation may give useful information about the search space. For this reason, in the first stages of evolution, the relaxation of the constraint violations may include some infeasible solutions in the population. The ε-constraint method has been widely used for single-objective optimization [17]. Recently, this method has gained a wide interest to solve CMOPs and has been integrated into variants of a decomposition-based algorithm which is MOEA/D [18]. Martínez and Coello [19] proposed the eMOEA/D-DE algorithm that uses a selection mechanism-based on a modified ε-constraint method. This new ε-constraint method was proposed to deal with the problem called the ε level comparison drift which occurs when the original ε-constraint method is used. The experimental results show that eMOEA/D-DE is highly competitive in dealing with CMOPs. However, for some test problems the computation of the ε-constraint value is not done in a proper way and the use of a misguided ε value will mislead the search of optimal solutions. Yang et al. [20] employed the ε-constrained method into MOEA/D-DE framework to form a new algorithm called MOEAD-$_\varepsilon$DE. In the experimental study, the authors compare MOEAD-$_\varepsilon$DE against four algorithms on CF-series test instances [21]. MOEA/D-$_\varepsilon$DE has the best performance in terms of IGD on CF1, CF6, and CF10. However, the results are preliminary and MOEA/D-$_\varepsilon$DE needs further improvements.

2.4 *Other Constraint Handling Techniques*

Most constraint handling studies have been developed for optimization problems consisting of only one objective. Researchers have provided detailed reviews of dealing

Table 1 Summary of some existing C-EMO works with various constraint handling techniques

References	Description	Test problems
Jiménez et al. [22]	Uses the Pareto concept to guide the feasible solutions towards the optimal front and the min-max formulation to guide the infeasible solutions towards the feasible region	CTP1 and CTP7, and OSY
Vieira et al. [23]	Transforms constraints into two objectives. The first objective is based on a penalty function, while the second objective is equal to the number of violated constraints	TBU, CPT6-CPT7, and OSY
Young [24]	Uses a combined value obtained by blending the individual's ranks in the ojective and constraint spaces	CTP6-CTP8
Geng et al. [25]	Adopts an infeasible elitist preservation mechanism to deal with CMOPs characterized with disconnected feasible regions and employs stochastic ranking to maintain the diversity	CONSTR, SRN, TNK, CTP1, CTP6, and CTP7
Oyama et al. [26]	Uses the idea of non-domination and niching concepts to solve CMOPs	Optimal design of a welded beam and conceptual design optimization of a two-stage-to-orbit spaceplane
Isaacs et al. [27]	Employs a non-dominated sorting for feasible and infeasible solutions. It transforms the CMOP into an unconstrained one by adding an additional objective which is the number of constraint violations	CTP2-CTP8
Ray et al. [28]	Keeps a small percentage of infeasible solutions during the evolutionary process in addition to the feasible solutions to search the feasible region	CTP2-CTP8
Liu and Wang [29]	Uses a constraint handling technique which is based on a temporary register. This strategy allows individuals with lower constraint violation values to participate in the crossover and mutation operators	CTP1-CTP8
Datta and Regis [30]	Uses surrogates for the objectives and constraints to determine the objective and constraint function values and a non-dominated sorting to find the most promising trial offspring solutions	BNH, SRN, TNK, OSY, ROBOT_2OBJ, ROBOT_3OBJ, ROBOT_5OBJ, MFG1, MFG2, BICOP1, BICOP2, TRICOP, G7, G18, G19

constraints with single objective optimization [31]. However, little is the effort that has been devoted to deal with CMOPs in the specialized literature. Fonseca and Fleming [32] proposed an algorithm that handles constraints by assigning a high priority to constraints and low priority to objective functions. Coello Coello and Christiansen [33] proposed to ignore the infeasible solutions. The implementation of this method is easy but it is difficult to find even a one feasible solution using this technique. This proof that an algorithm designed for C-EMO has to take into account various limits imposed on decision variables and convenient conceptualization of objective vectors. However, it is fundamental to take advantage of the profitable data supplied by infeasible individuals while dealing with C-EMO. Ray et al. [34] suggested the use of a Pareto ranking concept for both the objectives and the constraints. In fact, three non-domination rankings have been used in their work: (1) a ranking using the objective function values, (2) a ranking using different constraints, and (3) a ranking which is based on the combination of the objective functions and the constraints. The main advantage of using the Pareto ranking for both objectives and constraints is that it eliminates the problems of aggregation and scaling. Moreover, the method uses effective mating strategies which improve the convergence considerably. Harada et al. [35] designed a Pareto Descent Repair (PDR) operator that searches for feasible solutions out of infeasible individuals. Asafuddoula et al. [36] proposed to use an adaptive constraint handling scheme which is based on a violation threshold for comparison. In their constraint handling approach, a modified formulation of the constraint violation measure and a violation threshold measure were proposed. This method separates the constraint violation and objective function values and considers infeasible solutions with violations less than the identified threshold at par with feasible solutions. The experimental results on 10 benchmark CMOPs and on a real world submarine design problem [37] have shown the ability of this approach in dealing with CMOPs. However, the authors compared their approach to only NSGA-II. Hence, further comparisons are needed. Table 1 summarizes some of the existing C-EMO approaches in the literature. Recently, multi-objective optimization with a high number of objectives called many-objective optimization problems have gained a wide interest [18, 38]. In the literature, there are few works that have been proposed to solve constrained many-objective optimization problems [39, 40]. Thus, it will be interesting to design new constraint handling methods for constrained many-objective optimization problems.

3 Engineering Applications

Multi-objective optimization has always been an important area of application to which researchers have widely contributed [41–43]. The literature shows that multi-objective problems can be transformed into many-objective optimization and can be integrated with decision making tools for efficient productivity. The following (Tables 2 and 3) presents a brief outline of contributed research work in C-EMO.

Table 2 C-EMO applications in real world engineering optimization

Contributers	Application area
Kurpati et al. [44]	(a) Speed reducer design (minimize the volume, minimize the stress in one of the gear shaft)
	(b) Oil carrying fleet of ships (minimize the overall cost of building and operating a fleet of oil tankers), (maximize the cargo capacity of the fleet)
Aute et al. [45]	Air cooled condensing unit optimization (maximize heat rejection through the condenser coil and minimize the cost)
Pinto [46]	Supply chain management optimization (minimize cost (manufacturing, transportation) and maximize profit)
Sarker and Ray [47]	Crop rotation planning optimization (contribution maximization and cultivation cost minimization)
Chakraborty et al. [48]	Embedded system design optimization (maximize the performance and minimize the additional area requirement)
Sardiñas et al. [49]	Optimization of cutting parameters in turning processes (maximize tool life and maximize operation time)

As discussed earlier sections, C-EMO has wide spread applications in engineering science. In next subsections, we will elaborate each engineering applications and discuss how the C-EMO problem has been handled. This application will provide a base to apply such techniques in the deficient areas where C-EMO techniques are yet to be applied.

3.1 Speed Reducer Design and Design of Oil Carrying Fleet of Ships

Two constraint handling approaches are proposed in the study of [44]. The authors have developed four constraint handling techniques using Multi-objective Genetic Algorithm (MOGA). All these four techniques are developed on the basis of Constraint Handling by Narayanan and Azarm (CH-NA) [50]. Two engineering design problems; a speed reducer design and the design of a fleet of ships are solved with these four C-EMO techniques and CH-NA. The comparative study clearly showed that all four C-EMO techniques are better than CH-NA. In the first constraint handling technique, the authors emphasized on evaluating "Constraints First Objectives Next" model. In this model, constraints are evaluated to separate out feasible and infeasible individuals. In the second approach, infeasibility is considered to handle constraints. Third approach used the information of violated constraints. The last approach is the hybridization of all three approaches. As a result, the non-dominated solution from last strategy can find more uniformly distributed non-dominated solutions. The study clearly showed that "Constraint-First-Objective-Next" model outperformed

Table 3 C-EMO applications in real world engineering optimization (continued)

Contributors	Application area
Li et al. [53]	Power generation loading optimization (minimize fuel consumption, minimize emissions,minimize total cost and maximize output)
Guo et al. [54]	Aircraft landing schedule optimization (minimize total delay and total cost)
Moser and Sanaz [55]	Automotive Deployment Problem optimization (Data Transmission Reliability and minimize Communication Overhead)
Abu el Ala et al. [56]	Optimization of electric power system emission (minimize fuel cost and emission)
Tripathi and Chauhan [57]	Optimization of planetary gear train (minimization of surface fatigue life factor of gear and minimization of gear box volume)
Puisa and Streckwall [58]	Propeller optimization (Maximization of efficiency and Minimization of cavity volume)
Hajabdollahi et al. [59]	Optimization of compact heat exchanger (maximum effectiveness and the minimum total pressure drop)
Rajendra and Pratihar [60]	Gait Planning of Biped Robot optimization (maximize dynamic balance and minimize power consumption)
Liu and Bansal [61]	Optimize boiler combustion process (maximize flue gas temperature field value and maximize heat transfer rate)
Wang et al. [62]	Aircraft design for emission and cost reductions optimization (minimize total delay and minimize total cost)
Pandey et al. [63]	Topology optimization of compliant structures (minimize compliance, minimize maximum stress, minimize weight)
Sorkhabi et al. [64]	Energy-noise wind farm layout optimization (maximize the energy generation and minimize noise production)
Droandi and Gibertini [65]	Blade design of Tiltrotor aircraft optimization (maximize propulsive efficiency and hover Figure of Merit)
Datta et al. [66]	Robot gripper design optimization (minimize fluctuation of gripping force, minimize force transformation ratio)

"Objective-First-Constraint-Next" model. All the four C-EMO techniques work with few parameters. The authors concluded that amount of infeasibility and the number of violated constraints are vital for efficient performance of any C-EMO technique.

3.2 *Air Cooled Condensing Unit Optimization*

One of the prime component of the Refrigeration and Air-Conditioning system (RAC) is air cooled cross-flow heat exchangers. Its design severely affects the efficiency and performance of the entire system. The authors used the condenser model proposed by [51]. The objectives that are considered in [45] are to maximize the heat rejection through the condenser coil and minimize the overall cost. The constraints which limits the performance of both the objectives are refrigerant and air-side pressure drop, interrelationship between tube length and combined width of the fans placed on the cabinet, coil height, tube length. A GA based C-EMO coupled with a condenser simulation tool is developed to obtain the non-dominated solutions. Two multi-objective genetic algorithms are used to solve the C-EMO problem [32, 52]. The obtained solutions suggested to use fewer number of fans and tube of small lengths added with number of parallel circuits and more number of fins.

3.3 Supply Chain Management

Supply Chain Management (SCM) is concerned with integrating the efficient flow of materials, information, and finances as they move from supplier to consumer via manufacturer, wholesaler and retailer. The important benefits of SCM includes improved inventory management system, better sales forecasting, balanced supply-demand, elimination of irrelevant elements, improved balanced plans, working strategies and stronger partnerships. The author [46] has implemented NSGA-II in a hypothetical problem which is similar to usual problem encountered in real life. The important objective functions considered are manufacturing cost, transportation cost, profit. Many constraints, such as plant capacities, supplier capacities, inventory-balancing and total operating cost have been considered simultaneously. The study also concluded that EAs are highly efficient techniques for solving combinatorial problems.

3.4 Crop Rotation Planning Optimization

Reference [47] solved multi-objective crop planning problem. The crop planning problem is modeled as a linear and a non-linear bi-objective problem. A C-EMO is developed which is similar to NSGA-II with some modifications. The modifications is done in the parent selection and population reduction process. However, it is more computationally expansive than NSGA-II [14]. The hyperarea of the objective space is considered while selecting parent population from the feasible solutions in the same front. To maintain diversity in the variable space, the method not only maintain the extreme points but also considers minimum and maximum values of the decision variables. Few cases of bi-objective linear and nonlinear constrained optimization problems is solved with proposed Multi-objective Constrained Algorithm (MCA). Both algorithms (MCA and NSGA-II) have faced some difficulties to generate feasible solutions for the linear version of the problem. The comparative study shows that MCA outperforms NSGA-II for the linear version of the crop planning problem. The results of MCA are compared with NSGA-II. The study also discusses the sensitivity analysis of variables for non-dominated solutions.

3.5 *Embedded System Design Optimization*

The authors have targeted the embedded system design where the conflicting objectives are to maximize the performance and minimize the additional area requirement [48]. A hybrid MOGA is proposed which is a combination of multi-objective GA and multi-objective branch and bound method. The role of branch and bound method is to repair the infeasible solutions with the search progress. The comparative study between with or without repair clearly shows that the repair method is able to generate better set of non-dominated solutions.

3.6 *Optimization of Cutting Parameters in Turning Processes*

There are different cutting parameters in a turning process, such as cutting speed, feed and depth of cut, which are optimized as decision variables while the two conflicting objectives-maximization of tool life and the operation time is considered in the study of [49, 67]. Micro-GA [68] is used as the solution technique because of low computational time and wide spread diversity maintenance mechanism. It is a hard task to have the domain knowledge of machining, tool life empirical equations, surface finish, cutting forces, cutting power requirement to arrive at real life constraint of machining and to co-relate it with the actual performance desired by using mathematical and numerical optimization techniques. The tool life is severely affected by the Material Removal Rate (MRR) which is directly proportional to the tool wear and the cutting time. As the cutting time increases, the surface temperature also rises which decorates the surface finish. The authors also suggested that more number of conflicting objectives and constraints can be integrated to this problem. Such constraints or objectives could be surface temperature of tool, surface finish of workpiece, cost of machining, tool wear to name a few.

3.7 *Power Generation Loading Optimization*

In the advent of depletion of fossil fuels, the coal based power plants are facing acute shortage of fuel supply added with increased emissions of poisonous gases such as carbon monoxide, nitrogen oxides. Hence, in current scenario of power generation, it becomes essential to optimize the parameters such as minimum fuel consumption, maximum power output and minimize the emissions within the prescribed environmental limits. The authors proposed a constraint handling method incorporating the PSO algorithm for the power generation application. The same method will be applicable mainly to coal fired power plants but can also be extended to similar applications where the above three parameters need to be optimized simultaneously. The authors incorporated the dominance concept and used a few selection mechanism rules to guide the search direction from infeasible to feasible region. The method is independent of parameters that to be tuned by the user.

3.8 Aircraft Landing Schedule Optimization

The Air Traffic Control (ATC) on all the major and busy airports round the globe faces the difficulty of scheduling the flights landings on daily basis. The ATC has to schedule flight landing within a small time frame with sufficient separation between adjacent flight landing, such that total delay and total cost is minimized without compromising with the comfort and safety of the passengers. The authors [54] have focused for the development of a C-EMO problem that occurs in ATC. They have proposed two methods to handle constraints which is able to locate feasible region in a search space while NSGA-II and other first-come-first-served (FCFS) approach is inefficient in finding the feasible solutions.

3.9 Automotive Deployment Problem Optimization

The automotive industry installs Electronic Control Units (ECU) and data buses on vehicles of same type. There are approximately 50–80 ECUs which are individually connected to one of the 3–5 data buses. The software deployment problem has been formulated by the authors [55] in the form of bi-objective such that data transmission reliability is maximized and communication overhead is minimized. The reliability of data transmission is associated with functions such as precise actuation of air bag system based on the reliable data received from crash detection sensor to ECU which is passed on to airbag firing unit in quick succession such that overhead is minimum. The authors have used EA by incorporating the realistic constraints such as number of components on single ECU and size of ECU memory and compared the effects of diverse operators and constraint handling methods. The method consists of constraint handling mechanism coupled with repair mechanism.

3.10 Optimization of Electric Power System Emission

The main aim of Economic Power Dispatch (EPD) of electric power generation is to have optimal number of electricity generation facilities, which can meet the system load requirements, at the minimum possible fuel cost, satisfying all transmission and operational constraints including reduction in emission of toxic gases such as sulfur oxides and nitrogen oxides. The authors [56] developed a constrained differential evolution (DE) algorithm which is successfully tested and examined on the standard IEEE 30-bus test system.

3.11 *Optimization of Planetary Gear Train*

The Planetary gear train consists of two gears mounted so that the center of one gear revolves around the center of the other. This is used for applications such as 3D printing, automatic transmissions, etc. The multi-objective optimization of multistage planetary gear train is done by [57] which is a complicated problem due to the presence of integer variables. The two main contradicting objective functions include gear box volume and surface fatigue life factor which have been minimized by using Sequential Quadratic Programming (SQP) optimization technique and (NSGA II) [14]. It is observed that NSGA-II gives impressive results and concluded that planetary gear train occupies less volume and increases the efficiency of the system.

3.12 *Propeller Optimization*

A new constraint-handling technique for C-EMO is proposed based on behavioural memory technique (BMT) [58]. Initially, BMT was proposed for single objective optimization. The authors extended BMT technique for C-EMO by integrating with NSGA-II [14]. The multi-objective BMT is compared with an extra technique (a technique that in which summation of constraint violations are used) and constrained Pareto dominance technique. The comparative study showed that multi-objective BMT outperforms others.

3.13 *Optimization of Compact Heat Exchanger*

Due to the advancement in heat transfer technology and development of materials there has been tremendous improvement in the heat exchanger performance. The unique and distinctive feature of compact heat exchanger is a large heat transfer surface area per unit volume of the exchanger and they are widely used in various industries. The authors [59] have considered two conflicting objective functions in order to maximize the effectiveness and minimize total pressure drop for a triangular fin geometry with geometric constraints for a compact heat exchanger. The analysis has been carried out by integrating Computational Fluid Dynamics (CFD), Artificial Neural network (ANN) and NSGA-II [14] which has been named as CAN. The post optimal analysis concludes that the increase in heat transfer surface area does not have severe effect on the pressure drop.

3.14 *Gait Planning of Biped Robot*

Biped robot is a type of humanoid robot which resembles real human motions has been an active research topic in the field of bio-mechanics to understand the human body structure and behavior. The authors [60] have used multi-objective optimization in gait planning of a 7-dof biped robot while it ascends and descends staircases. The conflicting objective functions selected are maximum dynamic balance and minimum power consumption. Both GA and PSO Algorithm has been used to solve both constrained and unconstrained optimization problem. The study has also showed the performance of PSO is better than GA in terms of convergence, speed and diversity.

3.15 Optimize Boiler Combustion Process

The coal fired power plants are required to operate on high combustion efficiency with minimum carbon emission. But, the efficiency is greatly affected by slagging which occur if furnace exit gas temperature exceeds the fusion temperature of ash from the coal being fired and later deposits on furnace walls thereby reducing heat transfer. In this study authors [61] have integrated NSGA-II with Computational Fluid Dynamics (CFD) in order to optimize the boiler combustion process using the objective functions as the flue gas temperature field value and the heat transfer rate of water wall. The results are compared with conventional neural network based boiler optimization methods and the proposed method is better in terms of performance. The results maintains higher heat transfer rate of boiler as well as temperature in the vicinity of the boiler furnace within the ash melting temperature limit.

3.16 Aircraft Design for Emission and Cost Reductions

In the era of environmental protection, the aircraft which is the fastest mode of transportation, has to also deal with strict regulation of pollutants emissions which includes carbon dioxide, nitrogen oxides, hydrocarbons, smoke, soot and other gases. These gases not only responsible for polluting ground environment but also affects upper atmosphere. In this study, authors [62] have targeted to minimize emissions and cost using NSGA-II for constrained multi-objective optimization. A sensitive analysis is performed which shows that on reduction of cruise altitude and mach number results in lowering of global warming impact at the expense of operating cost.

3.17 Topology Optimization

The study focuses establishing a constructive solid geometry-based topology optimization (CSG-TOM) technique for the designing a compliant structure and mechanism using multi-objective optimization [63]. This method can deal with voids, non-design constraints, and irregular boundary shapes of the design domain, which are critical for any structural optimization. The constraints which restricts the search space are weight and volume fraction have been considered. NSGA-II is combined with finite element technique to handle constraints and to achieve non-dominated solutions. The authors compared the results with statc of the art techniques for 2-D and 3-D geometries.

3.18 Energy-Noise Wind Farm Layout Optimization

A constrained multi-objective energy-noise wind farm layout optimization is solved [64]. The constraints are handled with dynamic penalty functions. Thereafter, a hybrid approach is developed by combining a constraint Programming model with penalty functions to improve the objective functions. In the initial generations, the local search associated with the dynamic penalty approach used a smaller penalty factor. This method tries to achieve the closest feasible solutions near to the infeasible region. But a global search has been carried out by the penalty functions to penalize the objective functions of the infeasible region. It has been observed that the hybrid approach has better performance than penalty approach but affected by amount of feasible solutions corrected by the Constraint Programming model. The computational cost also increases with the introduction of local and global searches.

3.19 Blade Design of Tiltrotor Aircraft

Tiltrotor aircraft are used for vertical take-off and landing as they have tiltable rotors named as Proprotors and thus the capabilities of both helicopter and airplane. Such kind of versatility gives the advantage of cutting down fuel consumption during cruising and elevating the maximum cruising speed. However, the critical design challenges for such type of proprotors are bit complex. The authors [65] have considered three objective functions which includes maximization of propulsive efficiency during cruising at high speed, vertical climb and Maximization of the hover Figure of Merit. Here constrained NSGA-II [14] algorithm has been used for three objective optimization process coupled with a 2D-aerodynamic solver.

3.20 Robot Gripper Design Optimization

The authors have dealt with design optimization of a 7-link gripper [66]. The actuator model has been integrated into the robotic gripper problem. A generic actuation system (for e.g. a voice coil actuator) where force proportional to the applied voltage is considered. The design variables are the link lengths and the joint angles. Constraints that limits the search space are due to geometry and force requirements. The constraints are non-linear and multi-modal in nature. The bi-objective problem is solved using multi-objective evolutionary algorithm (MOEA) to optimally determine the dimensions of links and the joint angle of the robot gripper. The authors have developed a relationship between force and actuator displacement using the set of non-dominated solutions which can provide can provide the decision maker better insight to select the appropriate voltage and gripper design for appropriate application.

4 Medical Applications

Although C-EMO has been widely applied in engineering problems, use of medical application is rare and slowly gaining popularity. Some potential applications of C-EMO in medical science are described below.

4.1 Radiation Treatment Planning

Cancer is fast turning into an epidemic in the world, about more than twenty two million registered patients with most patients required to be treated with radiation therapy. Radiation therapy uses high-energy radiation to shrink tumors and kill cancer cells by damaging their DNA. Cancer cells whose DNA gets damaged ultimately stops dividing and dies. When the damaged cells die, they are broken down and eliminated by the body's natural processes. Radiation therapy can also damage normal cells leading to side-effects. Thus, amount of dose (energy deposited per unit mass) and area of exposure needs to be carefully selected to minimize the damage to health cells. The objectives of this radiation treatment planning comprises:- A high dose to be delivered (but below the critical value fixed for different cells) to the target area with a margin taking into account for position inaccuracies, patient movements, etc. for specific number of treatment over a set period of time [69]. Preventing the damage to healthy tissues and sensitive organs by minimizing the exposure time. This problem can be posed as C-EMO problem with many critical constraints.

4.2 Medical Image Processing

In order to inspect the vital internal organs of human body, an oncologist heavily rely on the information provided by various 3-D imaging techniques such as Computed Tomography (CT), Ultrasound, Magnetic Resonance Imaging (MRI), etc. These techniques combine the various projections obtained to create a 3D image. Thus for Medical Image Reconstruction for CT image 3 objectives were considered by [70] as to minimize the sum of the squared error between the original projection data and the re-projection data, Optimize the local smoothness in the neighborhood pixel of the reconstructed image and if the image is contaminated with noise, then maximize the entropy should be performed. This area also has a lot of opportunities for exploring the capabilities of C-EMO techniques.

4.3 Computer Aided Diagnosis

This is concerned with analyzing the medical data with the help of computer, for example a breast mammogram. In this, features extracted from both normal and diseased cases are compared to arrive at a conclusion by a classifier. The classifier considers two contradicting objective function namely sensitivity description (abnormal cases) and specificity description (normal cases) [71]. Researchers are encouraged to perform more deeper research in medical applications with C-EMO techniques to serve the humanity.

5 Conclusions

In this chapter, we have discussed about constraint handling in multi-objective evolutionary optimization techniques which have an added advantage over single objective optimization problems in context of efficient decision making. The real application of C-EMO can be found in various branches of engineering such as basic science, agricultural science, thermal science, electronics, robotics, operation management, etc. in which researchers have contributed. One of the potential area could be medical science. C-EMO techniques can be applied in radiotherapy treatment planning, brachytherapy, medical image processing, etc. Presently, researchers have started putting focus on multi-objective optimization with more than three objectives. Many-objective (more than 3 conflicting objectives) constraint handling will be a potential research direction. There are still many void areas of application which require attention of these techniques to attain optimized parameters for efficient and fruitful results.

References

1. Bechikh, S., Ben Said, L., Ghédira, K.: Negotiating decision makers' reference points for group preference-based evolutionary multi-objective optimization. In: 2011 11th International Conference on Hybrid Intelligent Systems (HIS), pp. 377–382. IEEE (2011)
2. Bechikh, S., Chaabani, A., Ben Said, L.: An efficient chemical reaction optimization algorithm for multiobjective optimization. IEEE Trans. Cybern. **45**(10), 2051–2064 (2015)
3. Bechikh, S., Kessentini, M., Said, L.B., Ghédira, K.: Chapter four-preference incorporation in evolutionary multiobjective optimization: a survey of the state-of-the-art. Adv. Comput. **98**, 141–207 (2015)

4. Azzouz, N., Bechikh, S., Ben Said, L.: Steady state ibea assisted by mlp neural networks for expensive multi-objective optimization problems. In: Proceedings of the 2014 Conference on Genetic and Evolutionary Computation, pp. 581–588. ACM (2014)
5. Mezura-Montes, E., Coello, C.A.C.: Constraint-handling in nature-inspired numerical optimization: past, present and future. Swarm Evol. Comput. **1**(4), 173–194 (2011)
6. Courant, R., et al.: Variational methods for the solution of problems of equilibrium and vibrations. Bull. Amer. Math. Soc **49**(1), 1–23 (1943)
7. Coello, C.A.C.: Theoretical and numerical constraint-handling techniques used with evolutionary algorithms: a survey of the state of the art. Comput. Methods Appl. Mech. Eng. **191**(11), 1245–1287 (2002)
8. Woldesenbet, Y.G., Yen, G.G., Tessema, B.G.: Constraint handling in multiobjective evolutionary optimization. IEEE Trans. Evol. Comput. **13**(3), 514–525 (2009)
9. Jan, M.A., Zhang, Q.: Moea/d for constrained multiobjective optimization: some preliminary experimental results. In: 2010 UK Workshop on Computational Intelligence (UKCI) (2010)
10. Li, H., Zhang, Q.: Multiobjective optimization problems with complicated pareto sets, moea/d and nsga-ii. IEEE Trans. Evol. Comput. **13**(2), 284–302 (2009)
11. Jan, M.A., Tairan, N., Khanum, R.A.: Threshold based dynamic and adaptive penalty functions for constrained multiobjective optimization. In: 2013 1st International Conference on Artificial Intelligence, Modelling and Simulation (AIMS), pp. 49–54. IEEE (2013)
12. Powell, D., Skolnick, M.M.: Using genetic algorithms in engineering design optimization with non-linear constraints. In: Proceedings of the 5th International Conference on Genetic Algorithms, pp. 424–431. Morgan Kaufmann Publishers Inc. (1993)
13. Deb, K.: An efficient constraint handling method for genetic algorithms. Comput. Methods Appl. Mech. Eng. **186**(2), 311–338 (2000)
14. Deb, K., Agrawal, S., Pratap, A., Meyarivan, T.: A fast and elitist multi-objective genetic algorithm: NSGA-II. IEEE Trans. Evol. Comput. **6**(2), 182–197 (2002)
15. Pal, S., Qu, B.Y., Das, S., Suganthan, P.N.: Optimal synthesis of linear antenna arrays with multi-objective differential evolution. Prog. Electromagn. Res. B **21**, 87–111 (2010)
16. Takahama, T., Sakai, S., Iwane, N.: Constrained optimization by the ε constrained hybrid algorithm of particle swarm optimization and genetic algorithm. In: AI 2005: Advances in Artificial Intelligence, pp. 389–400 (2005)
17. Takahama, T., Sakai, S.: Constrained optimization by ε constrained differential evolution with dynamic ε-level control. In: Advances in Differential Evolution, pp.139–154. Springer (2008)
18. Zhang, Q., Li, H.: Moea/d: a multiobjective evolutionary algorithm based on decomposition. IEEE Trans. Evol. Comput. **11**(6), 712–731 (2007)
19. Martínez, S.Z., Coello, C.A.C.: A multi-objective evolutionary algorithm based on decomposition for constrained multi-objective optimization, pp. 429–436 (2014)
20. Yang, Z., Cai, X., Fan, Z.: Epsilon constrained method for constrained multiobjective optimization problems: some preliminary results. In: Proceedings of the 2014 Conference Companion on Genetic and Evolutionary Computation Companion, pp. 1181–1186. ACM (2014)
21. Zhang, Q., Zhou, A., Zhao, S., Suganthan, P.N., Liu, W., Tiwari, S.: Multiobjective optimization test instances for the cec 2009 special session and competition. University of Essex, Colchester, UK and Nanyang technological University, Singapore, special session on performance assessment of multi-objective optimization algorithms, Technical report, vol. 264 (2008)
22. Jiménez, F., Gómez-Skarmeta, A.F., Sánchez, G., Deb, K.: An evolutionary algorithm for constrained multi-objective optimization. In: Proceedings of the 2002 Congress on Evolutionary Computation, 2002. CEC 2002, vol. 2, pp. 1133–1138. IEEE (2002)
23. Vieira, D.A., Adriano, R.L., Vasconcelos, J.A., Krähenbühl, L.: Treating constraints as objectives in multiobjective optimization problems using niched pareto genetic algorithm. IEEE Trans. Magn. **40**(2), 1188–1191 (2004)
24. Young, N.: Blended ranking to cross infeasible regions in constrainedmultiobjective problems. In: International Conference on Computational Intelligence for Modelling, Control and Automation, 2005 and International Conference on Intelligent Agents, Web Technologies and Internet Commerce, vol. 2, pp. 191–196. IEEE (2005)

25. Geng, H., Zhang, M., Huang, L., Wang, X.: Infeasible elitists and stochastic ranking selection in constrained evolutionary multi-objective optimization. In: Simulated Evolution and Learning, pp. 336–344 (2006)
26. Oyama, A., Shimoyama, K., Fujii, K.: New constraint-handling method for multi-objective and multi-constraint evolutionary optimization. Trans. Jpn. Soc. Aeronaut. Sp. Sci. **50**(167), 56–62 (2007)
27. Isaacs, A., Ray, T., Smith, W.: Blessings of maintaining infeasible solutions for constrained multi-objective optimization problems. In: IEEE Congress on Evolutionary Computation, 2008. CEC 2008. (IEEE World Congress on Computational Intelligence), pp. 2780–2787. IEEE (2008)
28. Ray, T., Singh, H.K., Isaacs, A., Smith, W.: Infeasibility driven evolutionary algorithm for constrained optimization. In: Constraint-Handling in Evolutionary Optimization, pp. 145–165. Springer (2009)
29. Liu, H.-L., Wang, D.: A constrained multiobjective evolutionary algorithm based decomposition and temporary register. In: 2013 IEEE Congress on Evolutionary Computation (CEC), pp. 3058–3063. IEEE (2013)
30. Datta, R., Regis, R.G.: A surrogate-assisted evolution strategy for constrained multi-objective optimization. Expert Syst. Appl. **57**, 270–284 (2016)
31. Michalewicz, Z., Dasgupta, D., Le Riche, R.G., Schoenauer, M.: Evolutionary algorithms for constrained engineering problems. Comput. Ind. Eng. **30**(4), 851–870 (1996)
32. Fonseca, C.M., Fleming, P.J.: Multiobjective optimization and multiple constraint handling with evolutionary algorithms. i. a unified formulation. IEEE Trans. Syst. Man Cybern. Part A Syst. Hum. **28**(1), 26–37 (1998)
33. Coello Coello, C.A., Christiansen, A.D.: Moses: a multiobjective optimization tool for engineering design. Eng. Optim. **31**(3), 337–368 (1999)
34. Ray, T., Tai, K., Seow, C.: An evolutionary algorithm for multiobjective optimization. Eng. Optim. **33**(3), 399–424 (2001)
35. Harada, K., Sakuma, J., Ono, I., Kobayashi, S.: Constraint-handling method for multi-objective function optimization: Pareto descent repair operator. In: Evolutionary Multi-Criterion Optimization, pp. 156–170, Springer (2007)
36. Asafuddoula, M., Ray, T., Sarker, R., Alam, K.: An adaptive constraint handling approach embedded moea/d. In: 2012 IEEE Congress on Evolutionary Computation (CEC), pp. 1–8. IEEE (2012)
37. Alam, K., Ray, T., Anavatti, S.G.: Design of a toy submarine using underwater vehicle design optimization framework. In: 2011 IEEE Symposium on Computational Intelligence in Vehicles and Transportation Systems (CIVTS), pp. 23–29. IEEE (2011)
38. Deb, K., Jain, H.: An evolutionary many-objective optimization algorithm using reference-point-based nondominated sorting approach, part i: Solving problems with box constraints. IEEE Trans. Evol. Comput. **18**(4), 577–601 (2014)
39. Jain, H., Deb, K.: An evolutionary many-objective optimization algorithm using reference-point based nondominated sorting approach, part ii: handling constraints and extending to an adaptive approach. IEEE Trans. Evol. Comput. **18**(4), 602–622 (2014)
40. Li, K., Deb, K., Zhang, Q., Kwong, S.: An evolutionary many-objective optimization algorithm based on dominance and decomposition. IEEE Trans. Evol. Comput. **19**(5), 694–716 (2015)
41. Bechikh, S., Said, L.B., Ghédira, K.: Group preference based evolutionary multi-objective optimization with nonequally important decision makers: application to the portfolio selection problem. Int. J. Comput. Inf. Syst. Ind. Manag. Appl. **5**(278–288), 71 (2013)
42. Kalboussi, S., Bechikh, S., Kessentini, M., Said, L.B.: Preference-based many-objective evolutionary testing generates harder test cases for autonomous agents. In: Search Based Software Engineering, pp. 245–250. Springer (2013)
43. Bechikh, S.: Incorporating decision maker's preference information in evolutionary multi-objective optimization. Ph.D. thesis, University of Tunis, ISG-Tunis, Tunisia (2013)
44. Kurpati, A., Azarm, S., Wu, J.: Constraint handling improvements for multiobjective genetic algorithms. Struct. Multidiscip. Optim. **23**(3), 204–213 (2002)

45. Aute, V.C., Radermacher, R., Naduvath, M.V.: Constrained multi-objective optimization of a condenser coil using evolutionary algorithms (2004)
46. Pinto, E.G.: Supply chain optimization using multi-objective evolutionary algorithms, vol. 15 (2004). Accessed Dec 2014
47. Sarker, R., Ray, T.: Multiobjective evolutionary algorithms for solving constrained optimization problems. In: International Conference on Computational Intelligence for Modelling, Control and Automation, 2005 and International Conference on Intelligent Agents, Web Technologies and Internet Commerce, vol. 2, pp. 197–202. IEEE (2005)
48. Chakraborty, B., Chen, T., Mitra, T., Roychoudhury, A.: Handling constraints in multi-objective ga for embedded system design. In: 19th International Conference on VLSI Design, 2006. Held Jointly with 5th International Conference on Embedded Systems and Design, 6 pp. IEEE (2006)
49. Quiza Sardiñas, R., Rivas Santana, M., Alfonso Brindis, E.: Genetic algorithm-based multi-objective optimization of cutting parameters in turning processes. Eng. Appl. Artif. Intell. **19**(2), 127–133 (2006)
50. Narayanan, S., Azarm, S.: On improving multiobjective genetic algorithms for design optimization. Struct. Optim. **18**(2–3), 146–155 (1999)
51. Jiang, H., Aute, V., Radermacher, R.: A user-friendly simulation and optimization tool for design of coils. In: Ninth International Refrigeration and Air Conditioning Conference (2002)
52. Srinivasan, N., Deb, K.: Multi-objective function optimisation using non-dominated sorting genetic algorithm. Evol. Comp. **2**(3), 221–248 (1994)
53. Li, L., Li, X., Yu, X.: Power generation loading optimization using a multi-objective constraint-handling method via pso algorithm. In: 6th IEEE International Conference on Industrial Informatics, 2008. INDIN 2008, pp. 1632–1637, IEEE (2008)
54. Guo, Y., Cao, X., Zhang, J.: Multiobjective evolutionary algorithm with constraint handling for aircraft landing scheduling. In: IEEE Congress on Evolutionary Computation, 2008. CEC 2008. (IEEE World Congress on Computational Intelligence), pp. 3657–3662. IEEE (2008)
55. Moser, I., Mostaghim, S.: The automotive deployment problem: a practical application for constrained multiobjective evolutionary optimisation. In: 2010 IEEE Congress on Evolutionary Computation (CEC), pp. 1–8. IEEE (2010)
56. El Ela, A.A., Abido, M., Spea, S.R.: Differential evolution algorithm for emission constrained economic power dispatch problem. Electric Power Syst. Res. **80**(10), 1286–1292 (2010)
57. Tripathi, V.K., Chauhan, H.M.: Multi objective optimization of planetary gear train. In: Simulated Evolution and Learning, pp. 578–582. Springer (2010)
58. Puisa, R., Streckwall, H.: Prudent constraint-handling technique for multiobjective propeller optimisation. Optim. Eng. **12**(4), 657–680 (2011)
59. Hajabdollahi, H., Tahani, M., Fard, M.S.: CFD modeling and multi-objective optimization of compact heat exchanger using CAN method. Appl. Therm. Eng. **31**(14), 2597–2604 (2011)
60. Rajendra, R., Pratihar, D.: Multi-objective optimization in gait planning of biped robot using genetic algorithm and particle swarm optimization tool. J. Control Eng. Technol. **1**(2), 81–94 (2011)
61. Liu, X., Bansal, R.: Integrating multi-objective optimization with computational fluid dynamics to optimize boiler combustion process of a coal fired power plant. Appl. Energy **130**, 658–669 (2014)
62. Wang, Y., Yin, H., Zhang, S., Yu, X.: Multi-objective optimization of aircraft design for emission and cost reductions. Chin. J. Aeronaut. **27**(1), 52–58 (2014)
63. Pandey, A., Datta, R., Bhattacharya, B.: Topology optimization of compliant structures and mechanisms using constructive solid geometry for 2-d and 3-d applications. Soft Comput., 1–23 (2015)
64. Sorkhabi, S.Y.D., Romero, D.A., Beck, J.C., Amon, C.H.: Constrained multi-objective wind farm layout optimization: introducing a novel constraint handling approach based on constraint programming. In: ASME 2015 International Design Engineering Technical Conferences and Computers and Information in Engineering Conference, pp. V02AT03A031–V02AT03A031. American Society of Mechanical Engineers (2015)

65. Droandi, G., Gibertini, G.: Aerodynamic blade design with multi-objective optimization for a tiltrotor aircraft. Aircr. Eng. Aerosp. Technol. Int. J. **87**(1), 19–29 (2015)
66. Datta, R., Pradhan, S., Bhattacharya, B.: Analysis and design optimization of a robotic gripper using multiobjective genetic algorithm. IEEE Trans. Syst. Man Cybern. Syst. **46**(1), 16–26 (2016)
67. Deb, K., Datta, R.: Hybrid evolutionary multi-objective optimization and analysis of machining operations. Eng. Optim. **44**(6), 685–706 (2012)
68. Coello, C.A.C.C., Pulido, G.T.: A micro-genetic algorithm for multiobjective optimization. In: Evolutionary Multi-Criterion Optimization, pp. 126–140. Springer (2001)
69. Lahanas, M., Milickovic, N., Baltas, D., Zamboglou, N.: Application of multiobjective evolutionary algorithms for dose optimization problems in brachytherapy. In: Evolutionary Multi-Criterion Optimization, pp. 574–587. Springer (2001)
70. Li, X., Jiang, T., Evans, D.: Medical image reconstruction using a multi-objective genetic local search algorithm. Int. J. Comput. Math. **74**(3), 301–314 (2000)
71. Devroye, L., Györfi, L., Lugosi, G.: A Probabilistic Theory of Pattern Recognition, vol. 31. Springer Science & Business Media, New York (2013)

Printed in the United States
By Bookmasters